The God of Science

IFEANYI CHUKWUJAMA

This Book is available at:
www.amazon.com
www.kindle.com
www.Jesus-On.com

For permission to reprint or copy this book, please
contact the publisher:

Ifeanyi Chukwujama

Jesus-On Kingdom Ministries

394-396 Warren Street, Boston, MA 02119

Our email: JesusOnkm@gmail.com

Our website: www.Jesus-On.com

ISBN: 10: 1503166570

ISBN-13: 978-1503166578

PRINTED IN THE UNITED STATES OF AMERICA

DEDICATION

This book is dedicated to God for bringing me to the light;
and for opening me up to this great and wonderful
experience and filling my heart with His knowledge and
His understanding; and for showing me things I never
knew and making me understand them.
And also to the memory of my father Enoch for teaching
me that humility is the best human virtue, and my mother,
Victoria for her love, tireless care and industry.
And also to my wife, Ukamaka, for her special love and
support, and for making life's journey with me through
thick and thin, and emerging with me at the other end of
the tunnel.
And also to my children Ifeanyi Jr., Chiamaka, Ifenna and
Chukwunonso for their love, inspiration and insights. They
have all been a very important part of my life's journey,
and together with my wife, played vital roles in helping me
shape my thoughts and find meaning for a great majority
of the things revealed to me by God as I wrote this book.
And also to my extended family who loved and supported
me and helped bolster my sense of worth when I had
nothing; and to a countless others, who made
contributions to my life, big or small. Everything is
important, because together they determine our directions
in life and who we ultimately become.
And also to you, the reader, because you are important to
our Lord Jesus Christ. Experience Him and live!
Glory Be to God!

Ifeanyi Chukwujama

CONTENTS

ACKNOWLEDGMENTS

Our God is great. And our Lord Jesus Christ is great, because He is one with the Father. He is faithful in all He has promised. I am a living example of His unfading love, His great mercy, His abundant grace, and His unlimited providence. The Bible says that with Him, everything is possible. And that is literally the meaning of my name (*Ife anyi Chukwu*).

And He has fulfilled His promise to me by making this book a reality. He has not only given me a book, He has taught me life, and I would forever be grateful to Him. He gave us His assurance that nothing can separate us from His love and I am holding on to that promise, because it never fails. There is nothing that comes out of the mouth of God that does not happen as He says it, because HE IS GOD. I am glad I know Him.

I thank God for my children and the great joy they bring into my life. Their readiness to help provided me with great assistance in the writing of this book, especially Ifenna, who helped me decipher the meaning of Christianity; and Chukwunonso who helped me in the design of the cover page and in document formatting; and Chiamaka for her general encouragement, and Ifeanyi Jr for overall support and editing of the book.

My wife deserves kudos for receptively listening to hours and hours of revelations I received while writing this book, offering insights and showing support for the messages in the book, in spite of her fragile health as she recovers from a life-changing surgery.

PREFACE

Christianity is not a religion. Christianity is **"Christ in us,"** because Jesus Christ comes into us and dwells in us when we accept Him into our lives. And by Christianity, I mean, establishing a relationship with, and having fellowship with Jesus Christ—that is, in case Christianity means other things to other people.

Jesus Christ did not call His followers Christians nor did His apostles and disciples call themselves that. But in the world today, we are known as Christians and we are proud to be; and our faith in Christ is known as Christianity, because we believe in Jesus Christ. And by coming to Jesus Christ, we receive life. He said to the Pharisees, *"You diligently study the Scriptures because you think that by them you possess eternal life. These are the Scriptures that testify about me, yet you refuse to come to me to have life." (John 5:39-40).*

Christianity is LIFE, because Jesus Christ is Life and comes to live in us when we accept Him as our Lord and Savior. And He told us in the Bible "I come that you may have life, and have it more abundantly." So when you come to Christ, you've come to LIFE. And you'll have life more abundantly. How then did we get from life to religion? Just like the Israelites before us, we took the perfect gift of God and turned it into man-made activities, designed more to gain glory for ourselves than to give glory to God. No wonder the world is getting darker and darker.

Religion is not about God. Religion is an attempt to imitate God. But nobody should imitate God for God detests such behavior. In the Book of Isaiah He says, *"I am the first and I am the Last; apart from me there is no God. Who then is like me? ..."* *(Isaiah 44:6-7)*. And He had decreed ruin for entities that are set up to imitate Him, and for those who mislead innocent people through their deceptive practices.

God put us in this world to reign, but the only way to reign is by serving Him. That is why King Solomon said: *"Of making many books there is no end, and much study wearies the body." "Fear God and keep his commandments, for this is the whole duty of man."* *(Ecclesiastes 12:12-13)*. I have finally embraced that truth and the result is abundant life.

Learning gives you knowledge. God gives you wisdom. When both are combined the way they should, you gain understanding. With much knowledge and no wisdom, you are dangerous even to yourself. With wisdom and no knowledge, you are ignorant, but the advantage is that your wisdom tells you when you must get knowledge and helps you to get it. Knowledge by itself can never lead you to wisdom, because wisdom only comes from God and is given only to the humble at heart.

From now on, I am staying right here with our

Lord Jesus Christ. If he stirs me left, I will go left. If he stirs me right, I will go right. He has finally shown me who I am supposed to be, and I am thankful to Him for that.

My goal is to help anyone who is interested in finding joy in life to find it, as I have found it, by sharing the message of the gospel with them the way it was given to me by our Lord Jesus Christ.

Giving the message to others so they can share in the joy of Jesus Christ is a responsibility that all believers have. Here is what the Bible says: ***"How, then, can they call on the one they have not believed in? And how can they believe in the one of whom they have not heard? And how can they hear without someone preaching to them?"*** *(Romans 10:14).*

*"**And how can they preach unless they are sent**? As it is written, 'How beautiful are the feet of those who bring good news!' But not all of the Israelites accepted the good news. For Isaiah says, 'Lord, who has believed our message?'"* *(Romans 10:15-16).*

*"**Consequently, faith comes from hearing the message, and the message is heard through the word of Christ.**"* *(Romans 10:17).*

Christ's goal is to build both the Jews and the Gentiles into one new man, built up together to become a holy temple to the Lord. If you have not reached out to the Jews, you have grounds to cover; and if you have not reached out to the Gentiles, you are trailing on the great commission and need to get started for the end is near.

To make it easier for a non-believer to know Jesus Christ, the question should be, *"Do you know that Jesus Christ is the reason you are alive and operational today, and do you know that He died on the cross so you 'may have life and have it more abundantly?'"*

Apostle Paul said to the Corinthians, *"Brothers, I could not address you as spiritual but as worldly — mere infants in Christ. I gave you milk, not solid food, for you were not yet ready for it. Indeed, you are still not ready for it. For since there is jealousy and quarreling among you, are you not worldly? Are you not acting like mere men? (1 Corinthians 3:1-3).*

And he also said: *"The man without the Spirit does not accept the things that come from the Spirit of God, for they are foolishness to him, and he cannot understand them, because they are spiritually discerned."* (1 Corinthians 2:14).

Anyone new to the gospel is just getting started and is simply an infant in Christ, as apostle Paul described in the passages above. As such, he should be given milk, not solid food, as Paul also stated. It is easier to anyone not knowledgeable about the gospel to understand abundant life than to understand eternal life. Jesus Christ came to give us both.

Getting people started with something they can relate to in their current lives is more effective than starting them out with a deeper spiritual thing as eternal life. Everybody is already living and understands one or two things about life and all of its difficulties: lack; sickness; making bad decisions and facing their consequences; sadness and sorrows and so on. And everyone would like to have an easier time dealing with these difficulties. As a result, people are more readily interested in finding a life they could manage better, as opposed to discussing issues they do not have the slightest understanding of, nor would, after being witnessed to just once. Most people in the world today are interested in how they can get some relief from life's difficulties, and they will find it in Christ.

Consider how Jesus Christ treated the Samaritan woman at the well. He asked her for water (everyday life's issue) and quickly let her know He could give her better water; the kind that could take away her thirst forever. And she immediately asked to receive it because she knew it is better than what she had been getting all her life. And she brought others to Christ. That is the power of the gospel. It is that captivating. And it is only through serving Christ that abundant life comes. And anyone who has abundant life (not life's excesses) also has eternal life.

It is important for everyone who is called to the gospel of Jesus Christ to stay with the message, and remain humble. The gospel has no equal because God has no equal. The light does not compete with darkness so why would the God of the universe compete with the things He created? What Jesus

Christ has given to us is not a religion, but life, therefore, no religion competes with Christ. So we are clearly not in competition with anyone. Our job is to give the message and the Holy Spirit of God will make the message do what it is designed to do. With that understanding, no one Christian is in competition with any other Christian; and no one Christian organization is in competition with any other Christian organization.

From apostle Peter's message of encouragement to the church about apostle Paul's writings to the churches, it is clear that Peter, who was made head over his brothers by Jesus Himself, was doing everything in his power to promote the gospel of Jesus Christ, as received and articulated by another apostle, who actually came much later to the calling. Peter wrote: *"Bear in mind that our Lord's patience means salvation, just as our dear brother Paul also wrote you with the wisdom that God gave him. He writes the same way in all his letters, speaking in them of these matters. His letters contain some things that are hard to understand, which ignorant and unstable people distort, as they do the other Scriptures, to their own destruction."* *(2 Peter 3:15-16).*

Apostle Paul also showed the same understanding in his address to the Corinthians; when they started fighting over whose style of delivery of the gospel to follow: Paul's, Apollos', Cephas'. *(I Corinthians 1; 2; & 3)*

Anyone who has the Spirit of God readily

recognizes messages that come from the Spirit of God, even when spoken by other people. That is why Eli the High Priest immediately recognized the message brought to him by the eleven-year-old Samuel who had to be taught by Eli that it was God who was trying to give Samuel a message.

Apostle Paul declared: *"So neither he who plants or he who waters is anything, but only God, who makes things grow. The man who plants and the man who waters have one purpose, and each will be rewarded according to his own labor."* (1 Corinthians 3:7-8).

In paraphrasing, the above declaration may then read:

"So neither" the Roman Catholic Church, nor the Apostolic Church, nor the Baptist Church, nor the Episcopal Church, nor the Methodist Church, nor the Lutheran Church, nor the Anglican Church, nor the Church of Christ … is anything, *"but only God, who makes things grow."* If the people preaching the gospel in all these churches maintain that Christ is the sole object of salvation; and that salvation can only be received though faith in Jesus Christ and the blood He shed on the cross, and in God who raised Him from the dead; they all *"have one purpose and each will be rewarded according to his own labor."* (I Corinthians 3:7-8)

With that understanding, nothing should really stand in the way of all the churches uniting, except the one thing we all deny but is actually the main reason why we cannot all get along: the wealth of the churches. The wealth that is involved in ministry has been the biggest obstacle that prevents the churches from uniting.

But it should not, if we separate our ambitions from our service to Jesus Christ. Instead of spending more time and energy promoting our individual brands of the gospel for various reasons, perceived or real, we should focus on jointly promoting the message of the gospel for the advancement of the kingdom of God, just like the apostles and the early disciples of Jesus Christ did.

Let us all remember the following Scripture: *"**But each one should be careful how he builds**. For no one can lay any foundation other than the one already laid, which is Jesus Christ. If any man builds on this foundation using gold, silver, costly stones, wood, hay or straw, his work will be shown for what it is, because the Day will bring it to light. It will be revealed with fire, and the fire will test the quality of each man's work. (1 Corinthians 3:10-13).*
*"**If what he has built survives, he will receive his reward. If it is burned up, he will suffer loss; he himself will be saved, but only as one escaping through the flames**." (1 Corinthians 3:14-15)*

Chapter 1

WHO IS GOD!

The only thing that is absolute in the universe is God. Everything else is relative, including the earth, the sun, the moon, man, nature and the entire cosmos. *(Genesis 1:1)*. Anything that is true is true because of God. What we see and experience appear the way we know them because there is God Almighty. If you look at anything and do not see God, then you have missed the whole point of the Bible and the meaning of life itself.

God must be the reference point for every consideration in this world, or nothing real could be achieved. He is everything and anything. His existence causes all else to exist. Without His existence, and continued presence in everything, nothing exists. *(Colossians 1:16)*. It all becomes a figure of anyone's imagination.

"I AM" is His name because He is, has been and will always be. *(Exodus 3:14)*. And because He is, we are, and everything else is. He is everybody and everything, and nothing exists outside of Him. He was. He is. He has always been. And He will forever be. *(Revelation 1:8 & Revelation 22:13)*.

He is the only truth, and there is no truth outside of Him. Truth is finite, not relative, because God is not relative. He decides what is true. He establishes what is true. He pronounces what is true because He designs truth, makes it and causes it to be. Nobody else knows it but Him because that is His being — The Truth. And whatever He says is, is. His Word is a decree and once He pronounces it; nothing changes it.

There is nothing outside of Him. He is nature. He is us. He is the entire creation, because if He is removed from it all, it will all cease to be. He is the beginning. He is the end. He is also the middle and everything in-between. If He isn't, nothing is. And there is nothing to compare Him to, because without Him, there is nothing to start with.

He is science. He is Mathematics. He is Physics, Chemistry and Biology. He is Law. He is Astronomy. He is Theology. He is the mountain. He is the sea. He is the sun, the stars and the moon. He is the wind and the calming breeze. He is the earth and the universe. He is all there is and nothing is, unless He is.

He is the argument and the counter argument. He is man's intelligence. And without Him, there is no man but an unsightly pile of organic mass. Not even the pile of organic mass would remain if He chooses not to keep it, otherwise, it would varnish altogether. Without Him, man would not have become a thought, much less the reality that he is today that continues to trouble God.

So let's stop the back and forth and start with "In the beginning, God". That is the only way we can find any true answers. It is the only cogent starting point for true science, true philosophy, true law, true astronomy, true engineering and any other field of study there is. HE IS GOD, and there is no other.

God is Spirit! He is God the Father; God the Son and God the Holy Spirit! That is why Jesus Christ said in the gospels: *"I and the Father are one."* *(John 10:30)* That, also, is why Jesus Christ said to His disciples:

"If you love me, keep my commands. ***16 And I will ask the Father, and he will give you another advocate to help you and be with you forever—*** *17* **the Spirit of truth**. *The world cannot accept him, because it neither sees him nor knows him. But you know him, for he lives with you and will be in you. 18 I will not leave you as orphans; I will come to you. 19 Before long, the world will not see me anymore, but you will see me. Because I live, you also will live. 20 On that day you will realize that I am in my Father, and you are in me, and I am in you. 21 Whoever has my commands and keeps them is the one who loves me. The one who loves me will be loved by my Father, and I too will love them and show myself to them."* *(John 14:15-21)*

The Spirit of God, for lack of a better analogy, is like the hologram. Every part of the whole has all the characteristics of the whole. Each additional part intensifies the effect of the concentration. The entire universe and all of the earth are not only completely filled with the Spirit of God; they and everything in them—living and non-living—are saturated with the Spirit of God. There is not a portion of anything visible or invisible on the earth, and in the entire universe, that is not saturated with the Spirit of God; and controlled by the Spirit of God.

This is the creation order as God laid it out in Genesis Chapter one. Anything that contradicts this order does not come from God; it is man-made and far away from the truth:

Day 1 of Creation (Genesis 1:3-5):

God brought forth the **light** and darkness receded. Light was not created on this day. Light was made to shine on the earth for the first time ever on this day. And God was the source of that light—*"God is light; in him there is no darkness at all" (*1 John 1:5); which makes sense because the Origin of light Himself (God) and the heavenly hosts who were present at the creation of the earth and the universe would not have been in darkness before this time. Therefore God has had light for all eternity because that is His nature. And God promises to be mankind's direct light again at the end of this age when the current universe passes away and the earth is renewed, Revelation 22:5.

God shined on the earth from Day 1 to Day 3, through His Son Jesus Christ (John 1:1-5). And since God made the earth spherical, the light illuminated one half of the sphere that is the earth while the other half was in darkness. The earth started to rotate so that every portion of it would receive the light within a 24-hour period. The appearance of light marked **noontime** on that first day. The **evening** came and **morning** followed, thus completing a full **day**—the first day.

The earth's gases started to form through ionization by the True Light. The Bible says that God is a consuming fire (Hebrews 12:29—*"for our God is a consuming fire"*). Therefore no celestial body has the ionizing power that equals God's—God is infinite energy plus the sum total of all the energies that exist everywhere in the universe, known and unknown, visible and invisible.

The dawn of light on the earth marked the beginning of **time** in the universe since the earth was created before time, and the universe was created on Day 4 of creation.

Day 2 of Creation (Genesis 1:6-8):

God created a void inside the water which totally encapsulated the earth and some water (Genesis 1:6). This is a spherical void; and this void separated water underneath the void from water on top of the void. God "stretched out" the void to form the current-day space (Genesis 1:7 & Isaiah 45:12—"*I have made the earth, and created man upon it: I, even my hands, have stretched out the heavens, and all their host have I commanded."*). God called this space heaven(s) (Genesis 1:8). This space at this point was under vacuum since it was entirely created inside water. And the water on top of the "firmament" was pushed out of the universe; still surrounds the universe till this today.

Day 3 of Creation (Genesis 1:9-13):

At God's command **(Genesis 1:9)**, the following events took place: Land came out of the water through volcanic eruptions. Water channels got trapped inside the earth, cooling the earth and forming water reservoirs to feed rivers and streams (Genesis 2:5-6).

Tsunamis ensued, violently clapping over the new land—cooling it, wetting it and weathering it. As land formed, water was pushed to one side to form one huge sea.

{When God destroyed the earth in the days of Noah, He had opened the earth and allowed some of the trapped water to come upon the earth's surface and create great flooding and devastated the earth. In Noah's flood, the earth's crust severally ruptured to let trapped water out to the surface of the earth; causing the land to break up into huge segments.}

{These segments later moved away from one another to form the various continents—the continental drift—*Genesis 10:25—**"One was named Peleg, because <u>in his time the earth was divided</u> ..."***Peleg was the 5th generation from Noah—born 201 years after the flood.}

At God's command (Genesis 1:11), the land that was formed, cooled and weathered earlier in the day produced vegetation (Genesis 1:12): The land produced grass and herb yielding seeds, and the fruit trees, whose seed was in itself, all after their own kinds.

{We should all remember that God made a plant grew overnight and produced enough foliage to provide shade for Jonah to stay under while Jonah was protesting God's leniency with the people of Nineveh. The plant then died by dawn the next day, arousing greater anger in Jonah (*Jonah 4:10—"It sprang up overnight and died overnight."*)}

The earth's atmosphere was completed between Day 1 and Day 3, since on Day 3 God created plants on the earth which depends on minerals from watered earth; gases from the atmosphere; and light energy from the light—the True Light that dawn on Day 1 and lasted to Day 3.

The Garden of Eden was planted by God on Day 3 to accommodate and provide for the man and the woman God would make on Day 6 of creation. (Genesis 2:8-14)—God planted all vegetation on Day 3.

The formation and configuration of the earth, its seas, its atmospheres and everything contained within them is complete by the end of Day 3, and sealed off by the Spirit of God, in preparation for the Big Bang that follows on Day 4.

Day 4 of Creation (Genesis 1:14-19):

God's command for lights in the expanse of heaven (Genesis 1:14-19) set off the Bing Bang, and trillions of oceans of light and fire flew in all directions across the space God had created on Day 2 (Genesis 1:6-8), filling it completely; aided by the vacuum God had created within space. {Space was created inside water—separating water from water—thereby drawing a vacuum; with the earth, its life, its seas and its atmosphere at the core of the space, held together and held in place by the Spirit of God (gravity)}.

The Spirit of God also protected the earth, its fragile vegetation and its atmosphere from being incinerated and completely consumed by the explosive energy of the Big Bang. This is God showing His infinite capacity to contain and protect whatever He wants to contain or protect. Through the Big Bang God created the universe and all the celestial bodies. But the Big Bang did not create life, the light, the earth, earth's seas (water), and earth's atmosphere (gases). And up to this day, there exists water outside the perimeters of—and completely surrounding—the current universe, according to Genesis 1:7. The firmament God called the heaven—and we today call space—separated water from water.

The fires and lights are imprints of the True Light which set off the Big Bang; and were arranged in clusters; set in orbits; contained by gravities; and given revolutionary, oscillatory and rotational motions to serve God's various purposes. The earth started to revolve around the sun to create the year. It also started to oscillate, 23 degrees to the south and 23 degrees to the north, to create seasons. And the universe as we know it today was born. God positioned the sun to take over from the True Light of Day 1 through Day 3 of creation as the source of light on the earth. He also created the moon to reflect the light from the sun and illuminate the night; and more importantly to serve as a sign pointing mankind to the sun and the stars being imprints of the True Light. God created the stars and completed an intricately balanced universe.

Day 5 of Creation (Genesis 1:20-23):

On Day 5 of creation, God created all sea animals and all the birds. He commanded for them and acted on His command (Genesis 1:21-22).

Day 6 of Creation (Genesis 1:24-31):

On Day 6 of creation, God created all land animals (Genesis 1: 24-25).

And finally, God created man. He created both man and woman on this Day 6 of creation: Adam was first and Eve followed later that day—*"So God created man in his own image, in the image of God created he him; male and female created he them. And God blessed them, and God said unto them, Be fruitful, and multiply, and replenish the earth, and subdue it: and have dominion over the fish of the sea, and over the fowl of the air, and over every living thing that moveth upon the earth. And God said, Behold, I have given you every herb bearing seed, which is upon the face of all the earth, and every tree, in the which is the fruit of a tree yielding seed; to you it shall be for meat. And to every beast of the earth, and to every fowl of the air, and to every thing that creepeth upon the earth, wherein there is life, I have given every green herb for meat: and it was so."* (Genesis 1:27-30).

"And God saw every thing that he had made, and, behold, it was very good. And the evening and the morning were the sixth day." (Genesis 1:31)

"Thus the heavens and the earth were completed in all their vast array." (Genesis 2:1)

Day 7 of Creation (Genesis 2:2-3):

"By the seventh day God had finished the work he had been doing; so on the seventh day he rested from all his work. ³ Then God blessed the seventh day and made it holy, because on it he rested from all the work of creating that he had done." (Genesis 2:2-3)

Study closely the following passage from the Bible, and the remarks I added in the brackets. God is telling the story of Genesis Chapter one, with wisdom at the foundation of it. The passage says that before God created anything—the earth or the universe—God created wisdom as the first of everything God created. Here is the passage:

"The LORD brought me forth as the first of his works,
before his deeds of old;
²³ I was formed long ages ago,
at the very beginning, when the world came to be. {YES! The earth began it all}
²⁴ When there were no watery depths, I was given birth, {starting with watery depths –Gen. 1:2}
when there were no springs overflowing with water; {Genesis 2:6 & Genesis 1:9}
²⁵ before the mountains were settled in place, {Genesis 1:9, Job 38:14}

before the hills, I was given birth,{Genesis 1:9 & Job 38:14}
²⁶ *before he made the world or its fields{Genesis 1:9}*
   or any of the dust of the earth.{Genesis 1:9}
²⁷ *I was there when he set the heavens in place,{Genesis 1:14-19}*
   when he marked out the horizon on the face of the deep,{Genesis 1:14-19}{The position and orientation of the light source in Genesis 1:3-5 is not recorded for us in the Bible}
²⁸ *when he established the clouds above{Genesis 1:2 & Job 38:9}*
   and fixed securely the fountains of the deep, {Genesis 1:9 & Job 38:8,10,11}
²⁹ *when he gave the sea its boundary{Genesis 1:9 & Job 38:8,10,11}*
   so the waters would not overstep his command, {Genesis 1:9 & Job 38:8,10,11}
and when he marked out the foundations of the earth. {Genesis 1:2 & Psalm 102:25 & Isaiah 48:13 }
³⁰   *Then I was constantly at his side. {Belongs with the TRINITY}*
I was filled with delight day after day, {Belongs with the TRINITY}
   rejoicing always in his presence, {Belongs with the TRINITY}
³¹ *rejoicing in his whole world{Belongs with the TRINITY}*
   and delighting in mankind. {Belongs with the TRINITY}

³² *"Now then, my children, listen to me; {Wisdom called mankind her children}*
   blessed are those who keep my ways.
³³ *Listen to my instruction and be wise;*
   do not disregard it.
³⁴ *Blessed are those who listen to me, {Mankind has to listen to her (wisdom) to prosper}*
   watching daily at my doors,
   waiting at my doorway.
³⁵ *For those who find me find life. {Wisdom leads to life}*
   and receive favor from the LORD. {Finding her guarantees God's favor}
³⁶ *But those who fail to find me harm themselves; {Rejecting wisdom brings pain and hurt to oneself}*
   all who hate me love death." (Proverbs 8: 22-36) {Rejecting wisdom is the final nail to anyone's coffin}

Summarizing, the passage says that God brought wisdom forth as the first of his works, before his deeds of old. It says that wisdom was formed very long ago, way before the physical world came into existence: at the very beginning, when the world came to be.

This passage is unequivocally declaring that the earth and its contents were created by God, before the universe was created. See how the passage went on describing the attributes of the earth, such as the mountains, the hills, the springs, the fields, etc. before it finally came to the creation of the heavens—our physical universe!

Then finally, in verse 27, the passage says: "*²⁷ I was there when he set the heavens in place,*
when he marked out the horizon on the face of the deep,
²⁸ when he established the clouds above
and fixed securely the fountains of the deep,
²⁹ when he gave the sea its boundary
so the waters would not overstep his command,
and when he marked out the foundations of the earth."

The word "heavens" in this passage stands for our physical universe. God created the universe on Day 4 of creation—*Genesis 1:14-19*—and filled the universe with signs and wonders to tickle man's imagination and get man to go looking for God. Unfortunate, however, when man's imagination was finally woken up, man went in every direction but the right direction. Because instead of being content with God's truth as God presented it to man in the Bible, man went in pursuit of his own theories and his own aggrandizement.

The proclamation, "I was formed long ages ago, at the very beginning, when the world came to be. When there were no watery depths, I was given birth *(Proverbs 8: 23-24)*;" alludes to the earth and the watery depth of Genesis 1:2 as the very beginning of the physical world and the physical universe.

It was after God created the heavens that God marked out the horizon on the face of the deep, established the clouds above the earth, and fixed securely the fountains of the deep. God established the clouds over the earth and the sea, and the fountains at the sea bed to circulate and regulate the earth's waters!

Ifeanyi Chukwujama

Chapter 2

THE SUPREMACY OF GOD

God is infinite energy plus the sum total of all the energies in the universe, visible and invisible; thermal, nuclear, sonic, light, magnetic, electrical, chemical, potential, hydrostatic, and all of the other forms of energy known and unknown:

$$E_{God} = E^{\infty} + \sum_{n=1}^{\infty} E_n$$

Where:

$E_{God} = God's\ Energy;$

$E^{\infty} = Infinite\ Energy;$

$E_n = Total\ Energy\ of\ its\ kind\ in\ the\ universe;$

n = number of energy types available in the universe, known and unknown; visible and invisible.

Jesus Christ *"ascended higher than all the heavens in order to fill the whole universe."* (Ephesians 4:10). And He *"is over all, through all and in all."* (Ephesians 4:6).

In essence, it is His presence in all things He

created that gives them their forms and functions; their energy, vitality and life (in the case of living things); and their utility and interconnectedness to one another. Jesus Christ came to the earth for all; to give eternal life to the entire mankind and to make their lives on earth more abundant.

God did not make everything and at the end of the sixth day walk away. He continues to operate within all things to maintain the cohesiveness and harmony He intended to achieve when He created everything. And for those who look at the account of creation in the Book of Genesis and doubt that God did what He said He did, you have not read the rest of the Bible, so you do not and cannot understand what God is saying in Genesis. You are welcome to study the Bible. It is God's gift to all humanity.

God created us and gave us great intelligence: He made us in His own image. So being as bright as we are, is not a feat we accomplished on our own; it is something that was given to us by God. The Bible says: *"But to each one of us grace has been given as Christ apportioned it. This is why it says: 'When he ascended on high, he led captives in his train and gave gifts to men.'"* *(Ephesians 4:7-8)*.

None of us was responsible for how much of that intelligence we received. It was all decided by God. He gave to each of us as He deemed necessary, not for us to become prideful but for the benefit of all humanity. That is why the Bible says: *"Each one should use whatever gift he has received to serve others, faithfully administering God's grace in its various forms."* *(1 Peter 4:10)*. And it also says: *"Now to each one the manifestation of the Spirit is given for the common good."* *(1 Corinthians 12:7)*.

24

God has no interest in justifying anything to anyone. He owes nothing to anyone. Rather, humanity owes Him for the elevated position we enjoy among all creation. The only demand God makes from any of us is faith. And if you cannot afford to give that much respect to Him, you have yourself to blame. Not only did humanity turn its back on God when Adam and Eve decided to disobey a simple command of God, we have continued to multiply the insult in so many different ways, including questioning the veracity of the things He said to us in the Bible.

Yet, out of His immense love for humanity, He sent His Son to suffer and pay for our sins; to redeem us and make us holy again, so we would be able to enter God's presence, which we have never truly known. This is a gift for all. Nobody needs to do anything to qualify for it. The only thing He requested from anybody to receive this gift is faith.

The faith we need to be cleansed of our sins and become acceptable to God is the same faith that we need to understand anything that comes out of the mouth of God. To understand the truth about anything in the universe, without assumptions and needless extrapolations, we need to have faith in God, and earnestly search the Scriptures for the truth about our world and the universe, at large.

And without faith, all our efforts continue to be a struggle – a matter of trial and error. We have done things that way throughout our history and we have not learned anything from our difficulties. The truth is glaring into everyone's face, and with the collective knowledge that exists in the world today, we could accelerate much faster in our understanding of the universe, with the added benefit of the forgiveness of our sins and the opportunity for eternal life.

When, with faith in God, you study the Bible, a new window of understanding is opened up to you. This new understanding completes and amplifies the understanding you already have in whatever specialty you are in. And things that never made sense to you before will begin to unfold right before your eyes. This is the revelation that is promised to anyone that has faith in God.

God does not discriminate. To knowledge, He adds knowledge. To wisdom, He adds wisdom. To understanding, He adds understanding. To hard work, He adds results. To humility, He adds greatness. To obedience, He adds honor and to faithfulness, He adds favor. Whatever you bring with you to the Bible, He refocuses and amplifies by the time you leave.

The opposite is also true of the Bible. If you come to the Bible with the intention of vilifying God, you will end up decimated and desecrated. Nobody can pick up a fight with God and succeed. If you came with doubts, that is a different thing. It is okay to have questions. And if you have questions and come with an open mind, you will satisfy your questions and more before you leave. God allows doubters to survive so they may have time to change their minds and gain the necessary knowledge they need to become enlightened.

To understand what God is telling you in Genesis about the creation of the universe and everything in it, you need to read the rest of the Bible in faith and in humility. The Bible is not just another book. It is the word of the all-powerful God of the universe, who does not play around but does everything with purpose. When you open the Bible, you are approaching the God that is the sum total of all the powers that exist in the universe and more. So you should approach reverently with thoughtfulness and eagerness to learn. That you are able to learn anything and retain it is Him.

The following passage from the Epistle of Peter testifies to the patience and love of God: *"But do not forget this one thing, dear friends: <u>With the Lord a day is like a thousand years, and a thousand years are like a day.</u> The Lord is not slow in keeping his promise, as some understand slowness. He is patient with you, <u>not wanting anyone to perish</u>, but everyone to come to repentance."* *(2 Peter 3:8-9)*. It also demonstrates that the Almighty God, who is above time and space, looks at everything whatever way He chooses. He does not count like we do because He is not constrained by the things we are constrained by.

So when you look at the timeline in Genesis and try to determine whether to believe God's account of the creation or reject it in light of current scientific projections, you will not only hamper yourself from gaining true understanding of things, you will also hurt your chance of salvation. To understand what God is saying to you as you read the Bible, you have to abandon your constrained methods of testing knowledge, and allow the Holy Spirit of God to take you through the information and reveal the hidden secrets to you.

Remember, God has put a spirit in you, which extends from His Spirit. When you allow your spirit to guide you as you delve into the deluge of God's wisdom that is the Bible, you emerge glistening with radiance and depth; having an understanding that negates everything else you have learned in your entire life. True knowledge comes from God, and He gives it to those who truly seek it.

The most important thing to remember is that the Bible was not given to us primarily to extend our knowledge of the universe; but to help us develop faith in God through His Son Jesus Christ, in whom are hidden all the treasures of wisdom and knowledge. Once you develop that faith, God, then, opens your eyes so you may see anything else you are looking for in the Bible. No matter what your situations or circumstances are, you can find the answers in the Bible. God is still speaking to anyone who would listen.

Here is the passage in the Bible that talks about Christ being the source of wisdom and knowledge: *"My purpose is that they may be encouraged in heart and united in love, so that they may have the full riches of complete understanding, in order that they may know the mystery of God, namely, Christ, in whom are hidden all the treasures of wisdom and knowledge." (Colossians 2:3).*

Note that apostle Paul calls it a mystery: that all the treasures of wisdom and knowledge are hidden in Jesus Christ. That is why no one can solve the mysteries of this life and the world we live in through worldly knowledge, but only through spiritual discernment. By having faith in Jesus Christ, and studying the word of God in the Bible, you will attain the wisdom and knowledge necessary for the fulfillment of life.

In essence, the Bible is also designed to give us knowledge and wisdom but only through understanding of Christ Jesus, because He is the embodiment of wisdom and knowledge. And in the very next verse, Apostle Paul reveals: *"I tell you this so that no one may deceive you by fine-sounding arguments."* *(Colossians 2:4)*.

The fine-sounding arguments include the so-called scientific information everybody is talking about these days. <u>Science will never contradict God</u> unless the information was erroneously collected. Science can only confirm the truth of God when properly managed. True science will only confirm the truth of God; and when anything called science contradicts God it is not science at all.

That something seems to hold some truth does not mean it is the truth. Truth is irrevocable and works all the time—not some of the time. The way the western civilization regard certain traditional practices and beliefs of people in under-developed cultures as superstition, is the same way Christianity must regard any science that contradicts God. Such science has no basis in the truth and would only mislead anyone who embraces it. The Theory of Evolution is one such example of superstitious science.

God created the universe and set all the natural laws in it. Here is the passage from the Bible: *"<u>If I have not established my covenant with day and night and fixed laws of heaven and earth</u>, then I will reject the descendants of Jacob and David my servant and will not choose one of his sons to rule over the descendants of Abraham, Isaac and Jacob. For I will restore their fortunes and have compassion on them."* *(Jeremiah 33:25)*. So it was God who set all the natural laws.

Therefore, whatever we are looking at in science and astronomy is nothing that God Himself did not create; set a purpose for; set limits, controls and uses for, according to His express Will. It is this same God who gave us the Bible, and promises us through the Bible, that whatever we ask for will be given to us. But can anyone ask for something he does not know he can receive? No! *(Romans 10:14)*.

So to ask and be sure to receive, you must have faith in the person you are making the request from, that firstly, he has the ability to do what you are asking for; and secondly, that he has the willingness to do what you are asking for. Only those who believe in Jesus Christ can make this kind of request and hope to receive what they ask for.

If anyone comes to Jesus Christ in faith and genuinely seeks information in any of the different specialty fields—science and astronomy or any other field of knowledge—He will generously reward the person. God puts clues to everything in life in the Bible, so the Bible should be a good starting point for all our investigations, if we hope to get it right.

God is really big. There is nothing in the universe that is outside of the reach and control of God. He made them all; He is in all of them, helping them keep their forms and functions, and their places in relation to one another.

It is not science that is alienating Christians as much as it is Christians alienating science. Science is the truth about nature. And God created nature and set all of its rules. Science simply catalogues what God has set up around us so we can have stability and assurance; and so we can ascertain for ourselves that God's words are indisputable. Therefore, real science supports everything God. And science should affirm the theology which adheres to the truth of the Scriptures; and not contradict it.

Most of what the world accepts as science is riddled with assumptions, extrapolations and compromises reached for lack of evidence or empirical data, in order to allow scientists to carry on with further investigations—especially since these investigations lead to the development of useful technologies and products that the world can use. That is why the world's scientific community receives and incorporates numerous revisions and corrections into science each year to get their conclusions closer to the truth.

God who created the universe and everything in it is the very one who provided humanity with the first hint of anything we know in science today. We can only know as much as He allows us to know. And if we force ourselves to figure out something God does not want us to know, He will confuse our minds so that we can never know it. It is God who stirs anybody's curiosity and sets people on a quest to figure something out.

All that is known in science and astronomy (and every field of study) to date was made possible by God. He is the Giver of knowledge and wisdom and understanding.

Our God intervenes whenever He chooses. His intervention alone can determine the course of any event. Yes! God has given each of us a destiny. Yet, depending on the choices we make in our day to day activities, He can alter events in our lives, leading to entirely different outcomes than was originally planned for any of us. The only way to achieve our destiny in this life is through faith in God. Consider the following passage from the Bible:

"The God who made the world and everything in it is the Lord of heaven and earth and does not live in temples built by human hands. [25] And he is not served by human hands, as if he needed anything. Rather, he himself gives everyone life and breath and everything else. [26] From one man he made all the nations, that they should inhabit the whole earth; and he marked out their appointed times in history and the boundaries of their lands. [27] God did this so that they would seek him and perhaps reach out for him and find him, though he is not far from any one of us. [28] 'For in him we live and move and have our being.' As some of your own poets have said, 'We are his offspring.'" (Acts 17:24-28).

The above passage says that *"In him we live and move and have our being,"* meaning that all we do in this world, we do in God; not outside of Him because He is everything and all things for us. That leads us to the conclusion that God is the sum total of all the energies that exist everywhere in the universe; enthalpies and entropies.

In other words, God is the sum total of all the energies in all the stars and planets in all the galaxies; plus all the energies in all the black holes and quasars; plus all the energies in the dark matters; plus the dark energy; and the energies in all of creation not known to man; plus infinite energy.

This conclusion is amplified in this passage from the Bible: *"God ascended higher than all the heavens in order to fill the whole universe."* (Ephesians 4:10). And He *"is over all, through all and in all."* (Ephesians 4:6).

Chapter 3

GOD RETAINS THE MASTER KEY TO EVERYTHING

God either retains the 'master key' equivalence of everything He made available to man; or remains solely in control of it. He gave us life, but it is not resident in us but rather radiates from Him. He gave us intelligence, yet He controls and directs it. He gives us knowledge but determines how much we acquire, the quality of it, and its usefulness to our lives.

He gave us languages; yet He retains the master language. He gave us freedom to express our wills but closely keeps watch over all our thoughts and actions, and demands accountability from each and every one of us for every word we utter from our mouth. He gave us salvation through the blood of His Son Jesus Christ, but decided it must be received by faith alone *"so that no one can boast. For we are God's workmanship, <u>created in Christ Jesus</u> to do good works, which God prepared in advance for us to do. "* *(Ephesians 2:8-10).*

He gave us technologies, but it feels as though He had not given anything out, for His technology chest is still full. *"Do not deceive yourselves. If any one of you thinks he is wise by the standards of this age, he should become a 'fool' so that he may become wise. For the wisdom of this world is foolishness in God's sight. As it is written: 'He catches the wise in their craftiness'; and again, 'The Lord knows that the thoughts of the wise are futile.' So then, no more boasting about men."* *(1 Corinthians 3:18-21).*

Human Science has estimated that everything we have been able to observe and detect in the universe constitutes less than five percent (5%) of the universe. The other ninety five percent (95%) of the universe remains a mystery to us. Scientists know that the rest of the 95% is there because of its effect on the things we know and understand, yet it is elusive to us. The unknown 95% of the universe has been classified by science as dark energy and dark matter. And here is an internet excerpt about them:

Source of Information:

(http://science.nasa.gov/astrophysics/focus-areas/what-is-dark-energy/)

"Dark Energy, Dark Matter

In the early 1990's, one thing was fairly certain about the expansion of the Universe. It might have enough energy density to stop its expansion and recollapse, it might have so little energy density that it would never stop expanding, but gravity was certain to slow the expansion as time went on. Granted, the slowing had not been observed, but, theoretically, the Universe had to slow. The Universe is full of matter and the attractive force of gravity pulls all matter together. Then came 1998 and the Hubble Space Telescope (HST) observations of very distant supernovae that showed that, a long time ago, the Universe was actually expanding more slowly than it is today. So the expansion of the Universe has not been slowing due to gravity, as everyone thought, it has been accelerating. No one expected this, no one knew how to explain it. But something was causing it.

Eventually theorists came up with three sorts of explanations. Maybe it was a result of a long-discarded version of Einstein's theory of gravity, one that contained what was called a "cosmological constant." Maybe there was some strange kind of energy-fluid that filled space. Maybe there is something wrong with Einstein's theory of gravity and a new theory could include some kind of field that creates this cosmic acceleration. Theorists still don't know what the correct explanation is, but they have given the solution a name. It is called dark energy.

What Is Dark Energy?

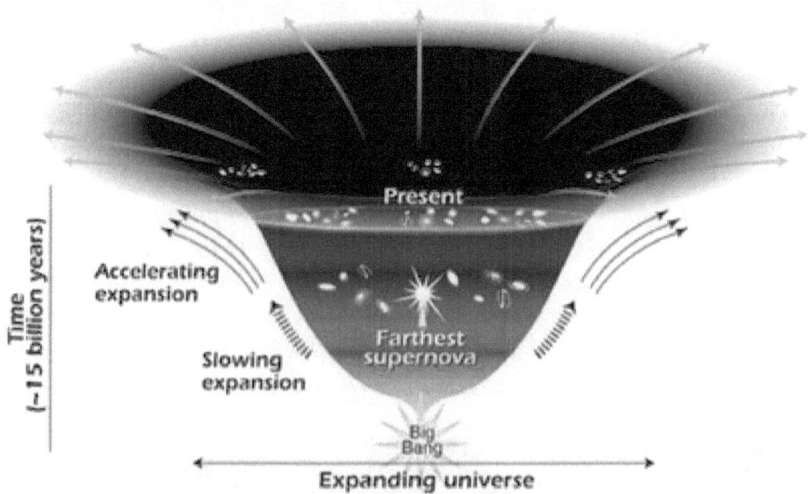

Universe Dark Energy-1 Expanding Universe

This diagram reveals changes in the rate of expansion since the universe's birth 15 billion years ago. The more shallow the curve, the faster the rate of expansion. The curve changes noticeably about 7.5 billion years ago, when objects in the universe began flying apart as a faster rate. Astronomers theorize that the faster expansion rate is due to a mysterious, dark force that is pulling galaxies apart.

NASA/STSci/Ann Feild

More is unknown than is known. We know how much dark energy there is because we know how it affects the Universe's expansion. Other than that, it is a complete mystery. But it is an important mystery. It turns out that <u>roughly 68%</u> of the Universe is dark energy. Dark matter makes up about 27%. The rest - everything on Earth, everything ever observed with all of our instruments, all normal matter - adds up to less than 5% of the Universe. Come to think of it, maybe it shouldn't be called "normal" matter at all, since it is such a small fraction of the Universe.

One explanation for dark energy is that it is a property of space. Albert Einstein was the first person to realize that empty space is not nothing. Space has amazing properties, many of which are just beginning to be understood. The first property that Einstein discovered is that it is possible for more space to come into existence. Then one version of Einstein's gravity theory, the version that contains a <u>cosmological constant</u>, makes a second prediction: "empty space" can possess its own energy. Because this energy is a property of space itself, it would not be diluted as space expands. As more space comes into existence, more of this energy-of-space would appear. As a result, this form of energy would cause the Universe to expand faster and faster. Unfortunately, no one understands why the cosmological constant should even be there, much less why it would have exactly the right value to cause the observed acceleration of the Universe.

Dark Matter Core Defies Explanation

This image shows the distribution of dark matter, galaxies, and hot gas in the core of the merging galaxy cluster Abell 520. The result could present a challenge to basic theories of dark matter.

Another explanation for how space acquires energy comes from the quantum theory of matter. In this theory, "empty space" is actually full of temporary ("virtual") particles that continually form and then disappear. But when physicists tried to calculate how much energy this would give empty space, the answer came out wrong - wrong by a lot. The number came out 10^{120} times too big. That's a 1 with 120 zeros after it. It's hard to get an answer that bad. So the mystery continues.

Another explanation for dark energy is that it is a new kind of dynamical energy fluid or field, something that fills all of space but something whose effect on the expansion of the Universe is the opposite of that of matter and normal energy. Some theorists have named this "quintessence," after the fifth element of the Greek philosophers. But, if quintessence is the answer, we still don't know what it is like, what it interacts with, or why it exists. So the mystery continues.

A last possibility is that Einstein's theory of gravity is not correct. That would not only affect the expansion of the Universe, but it would also affect the way that normal matter in galaxies and clusters of galaxies behaved. This fact would provide a way to decide if the solution to the dark energy problem is a new gravity theory or not: we could observe how galaxies come together in clusters. But if it does turn out that a new theory of gravity is needed, what kind of theory would it be? How could it correctly describe the motion of the bodies in the Solar System, as Einstein's theory is known to do, and still give us the different prediction for the Universe that we need? There are candidate theories, but none are compelling. So the mystery continues.

The thing that is needed to decide between dark energy possibilities - a property of space, a new dynamic fluid, or a new theory of gravity - is more data, better data.

What Is Dark Matter?

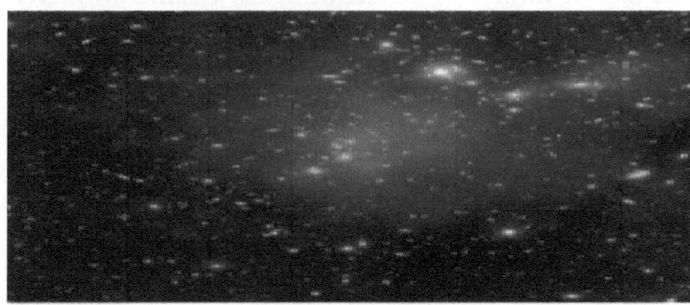

Abell 2744: Pandora's Cluster Revealed

One of the most complicated and dramatic collisions between galaxy clusters ever seen is captured in this new composite image of Abell

2744. The blue shows a map of the total mass concentration (mostly dark matter).

By fitting a theoretical model of the composition of the Universe to the combined set of cosmological observations, scientists have come up with the composition that we described above, ~68% dark energy, ~27% dark matter, ~5% normal matter. What is dark matter?

We are much more certain what dark matter is not than we are what it is. First, it is dark, meaning that it is not in the form of stars and planets that we see. Observations show that there is far too little visible matter in the Universe to make up the 27% required by the observations. Second, it is not in the form of dark clouds of normal matter, matter made up of particles called baryons. We know this because we would be able to detect baryonic clouds by their absorption of radiation passing through them. Third, dark matter is not antimatter, because we do not see the unique gamma rays that are produced when antimatter annihilates with matter. Finally, we can rule out large galaxy-sized black holes on the basis of how many gravitational lenses we see. High concentrations of matter bend light passing near them from objects further away, but we do not see enough lensing events to suggest that such objects to make up the required 25% dark matter contribution.

However, at this point, there are still a few dark matter possibilities that are viable. Baryonic matter could still make up the dark matter if it were all tied up in brown dwarfs or in small, dense chunks of heavy elements. These possibilities are known as massive compact halo objects, or "MACHOs". But the most common view is that dark matter is not baryonic at all, but that it is made up of other, more exotic particles like axions or WIMPS (Weakly Interacting Massive Particles)."

And what the scientists are discussing in the excerpt above is only the perceivable physical universe, and does not include the supernatural realm.

In regard to language, the whole world spoke the same language until the descendants of Noah refused to disperse and populate the face of the earth as instructed by God, but instead started to build a tower to reach heaven, so God confused them with different languages so they could no longer understand one another *(Genesis 11:1-9)*. They abandoned the tower project and went their separate ways. And different languages within different geographical areas led to different customs and traditions.

It can be assumed that the original earth language is the same as that spoken in heaven since God's original intention for man was to have people on earth do things as they are done in heaven. *(Matthew 6:10)*. So language is most certainly one of those things God intended to establish on earth as it is in heaven. We can never know for sure, nor is it really relevant to our lives today.

God knew that man was not going to perform as God intended, and He had already put a plan to rescue man in place — in the person of His Son Jesus Christ who would become man and come to earth to die for man's sins so man could receive salvation from God. And because of God's pre-knowledge of these events, man's first language on earth could have been something different from the heavenly language. One thing, however, is true as demonstrated in the Scripture:

The heavenly language is a master language, not spoken anywhere on the planet earth. This was demonstrated by the first Christians when they first received the Holy Spirit. *"All of them were filled with the Holy Spirit and began to speak in <u>other tongues</u> as the Spirit enabled them.*

"Now there were staying in Jerusalem God-fearing Jews from every nation under heaven. When they hear <u>this sound,</u> a crowd came together in bewilderment, because each one heard them speaking in his

own language. Utterly amazed, they asked: 'Are not all these men who are speaking Galileans?

"Then how is it that each of us hears them in his own native language? Parthians, Medes and Elamites; residents of Mesopotamia, Judea and Cappadocia, Pontus and Asia, Phrygia and Pamphylia, Egypt and the parts of Libya near Cyrene; visitors from Rome (both Jews and converts to Judaism); Cretans and Arabs — we hear them declaring the wonders of God in our own tongues!' Amazed and perplexed, they asked one another, 'What does this mean?'" (Acts 2:4-12)

That is the heavenly language — one tongue which everybody on earth and in heaven understands and would easily speak effortlessly once we get to heaven. On that day of Pentecost, the Holy Spirit descended on everyone in that room and they spoke in other tongues which together produced <u>the sound</u> that is understood by every listener is his own native tongue.

Just as God promises to write His Law on everyone's heart so that no one will need to teach anyone else about the Law of God; He will also write this heavenly language in everyone's heart so that we can speak it fluently and understand one another without the need for interpreters. In our current state, we could understand the heavenly language as evidenced in Acts Chapter 2.

No matter your native tongue, you can fully understand anything that God or any heavenly host says. And this will be showcased in full scale when Christ returns to the world a second time. The whole world will understand Him as He addresses everyone at the same time.

And when the Apostle Paul overheard the heavenly

conversation, he understood what was being said in his own native tongue (Hebrew) even though the conversation was in heavenly language — just as was demonstrated with the other Apostles in Acts chapter 2. Here is the passage from the Bible regarding Paul's experience with the heavenly language: *"I know a man in Christ who fourteen years ago was caught up to the third heaven. Whether it was in the body or out of the body I do not know — God knows.*

"And I know that this man — whether it was in the body or out of the body I do not know, but God knows —was caught up to paradise. He heard inexpressible things, things that man is not permitted to tell." (2 Corinthians 12:2-4).

Paul heard what was said and understood them clearly: He listened in on heavenly conversations and understood them in Hebrew. The same way, the visitors to Jerusalem at the first Pentecost following Christ's crucifixion heard and understood in their own languages what the apostles were prophesying under the control of the Holy Spirit.

At the second coming, Jesus Christ would not need to address any of us in our own native languages, for us to understand Him. He will speak to the whole world using the heavenly language and everyone will understand Him. And He would not use any technologies developed by man to announce His presence to the world on that day-of-judgment. He has it all under control.

Everything humanity has been able to develop so far does not have the capacity to achieve Christ's purpose on that Day of Judgment. He would not need our radio, our television, our internet or even our buildings to do any of the things He came for. God allowed the development of these technologies for our own use; and to give us a glimpse at the wonders He has put together. He has technologies far more sophisticated than humans have the mental capacity to comprehend or to contain, much less manage.

We need to also remember that when Christ comes back to the earth, it would not be like the first time. He came as a baby before, tiny and helpless and relied on human parents to navigate through life, until He started His mission in the world. This time around, He is coming as King of kings and Lord of lords. And He is coming in all His glory. Man's help is out of the equation, because Jesus Christ is coming to judge man — all men, the righteous and the unrighteous.

God did not save any human fabrications from the great flood, in Genesis, to get Noah and his family going again. God erased everything that was made by man in that flood. The only human fabrication that survived the flood, the ark, was dictated to Noah by God with amazing specificity and precision, and was abandoned by Noah atop Mount Ararat because its purpose was completed. Noah and family had to build everything in their lives from the scratch with things placed on earth by God Himself.

If Noah, who was only a man, does not need to rely on previously built man-made things, why would God Himself rely on anything that was made by the hands of sinful man? If He saved some of our technologies for His use on the day He came to judge us, is He still Almighty? That does not sound like the Almighty God who first put it all together simply by speaking a word.

On His second coming, when He speaks, His voice will resonate all over the world. When He motions His hands, flashes of light will streak across the sky for all to see. And each word out of His mouth will be heard and understood by people of every tongue everywhere in the world, and all at once. That is God's kind of technology. Like the sun, the technology that provides heat and light to half the earth at once to fuel people's thoughts and creativity while the other half of the earth is in darkness and in a state of rest and restfulness; and at the same time, having the same effects on the other planets and satellites.

The kind of technology that combines colors to form light and bring to life everything it touches! The kind of technology that extends one matter, light, into different forms and properties: from long wavelengths and short amplitudes (radio wave, microwave) that transmit signals; to short wavelengths and long amplitudes (X-rays and gamma rays) that ionize other matters; and everything else in-between. The kind of technology that causes vast nuclear furnaces (stars) to spin around one another and converge into black holes and quasars.

In addition to our technology being too inferior for the creator of the universe to rely on, a judge does not come into the courtroom and ask the person under trial to ready the courtroom for his judgment, or to provide equipment for the telecast of the procedure. The judge comes in ready to listen to arguments and deliver his judgment. He is in charge. It is his courtroom. Everything is already set up the way he wants it.

He simply goes to work; listening to the arguments

and delivering his verdicts. Then he rises and leaves the courtroom before anyone else does. It is his playground and the person under trial has no part in anything that happens in that courtroom. He does as he is commanded to, and only when he is permitted to do so. His help is not necessary at any point in the entire proceeding. He is on trial and his fate is being determined. The court has no use for him at that point in time.

Such is the situation with humanity when Jesus Christ shows up on earth for the second time. All the kings of the world would abdicate their thrones and run helter-skelter in awe of the King who just arrived. His entry would simply nullify and make dysfunctional all human technologies as a way of announcing His entry and His complete control of everything, because He is no ordinary king. When an asteroid enters the earth's atmosphere from out of space, it streaks across the sky and sets off a shockwave that paralyzes our electrical and electronic systems, if it is large enough.

Then, how much more disruption could the grandeur entrance of the creator of the entire universe cause to earth's infrastructure once He enters earth's atmosphere? Just the thought of it alone is paralyzing, if you have good enough imagination. Our only savior that day is His assurance that those who believe in Him, and are called by His name, are going to be spared. He actually instructed us to look towards that event with great anticipation and joy, because it is our day of salvation.

Jesus Christ would not need our help on that day. When He shows up, He takes total control as only He can. Yes the world is round, but the whole world will see Him at once, and His judgment would be handed out expeditiously to every single person that ever walked on the face of the earth, with Him breaking no sweat. He is that fast! He is that efficient! He is that powerful! And for the first time, humanity will get a glimpse of how truly powerful He is.

Unfortunately, no unbeliever will have the opportunity to change his mind at that point. That is why the Scripture says: *"We live by faith, not by sight."* *(2 Corinthians 5:7).* If anyone waits for this demonstration of His might to repent and believe in Him, he has already missed the opportunity to do so. We were supposed to believe by faith and not wait for proof. The only sure time to repent and believe in Him is now!

Chapter 4

EQUAL AND OPPOSITE FORCES

God created heaven and earth. Out of Love, He made earth specially to hold life. The Bible says: *"For this is what the Lord says — he who created the heavens, he is God; he who fashioned and made the earth, he founded it; <u>he did not create it to be empty, but formed it to be inhabited</u> — he says: 'I am the Lord, and there is no other.'" (Isaiah 45:18).*

In the Bible, God also says: *"<u>It is I who made the earth and created mankind upon it</u>. My own hands stretched out the heavens; I marshaled their starry hosts." (Isaiah 45:12).* God created some very powerful forces in the world and constrained them within limits to keep them from destroying mankind and other living things on earth. In the book of Job, He talked about a few examples in His rebuke of Job. *(Job 38, 39, 40 & 41)*

When you look around the earth, there is turbulence and there is serenity. But even the things that appear to be very tranquil have serious movements inside of them. Where there is one force, God created an opposing force to cancel it, thereby restoring balance. This balance creates stability and provides reassurance that what we see around us today will be there tomorrow.

The force of gravity is pulling everything on the surface of the earth downwards. That is why nobody lifts up into the air unless he propels himself into the air by some mechanism. Even at that, he comes back down as quickly as he got up unless the force that lifted him off the ground is sustained to keep him up or continue to lift him further away from the ground. As soon as the force is removed, the person or object falls down to the ground. That is due to gravity.

So while gravity is pulling us downward, the solid ground is keeping us from falling any deeper into the earth. The rigidity of the ground is asserting a force on every object on earth equal to the force of gravity that is acting on that object. As such people move freely along the surface of the earth and do not feel the effect of gravity on them because an equal and opposing force is exerted on them by the ground.

While some stable conditions are a result of two equal and opposing forces, some are just the correct level of one force as adjusted by God. Cold is not an opposing force to Heat. Cold is simply an absence of heat. Heat activates, excites and expands matters but cold deactivates, deadens and shrinks matters. Yet cold is not a discrete force. It is all heat or absence of it. Yet heat and cold are equal and opposite because we can feel both. We thrive in the ambient temperatures.

Similarly, darkness is not opposed to light. Darkness is simply absence of light, and instantly vanishes when light appears. But because they affect life on earth differently, we have come to regard them as two opposing forces, because each impacts us differently. And God intended them that way.

Even light, itself, is a composite of colors mixed together to impart visual impressions on all living things. As a composite, it bounces off of things and brings them to life. However, when separated into constituent colors, each color has a different effect on us: from hot and intense to cool and mellowing. But together, they produce delight, optimism and even excitement.

God made heaven. He also made hell. While heaven is for the faithful and the obedient, hell is for the unbeliever, the disobedient and the wicked. God resides in heaven with the heavenly hosts and the saints; and hell is designed for Satan and his gang of bad angels, and even humans who decided to become agents of evil, corrupted by Satan and his fallen angels. Every person that ever lived on earth is going to end up in heaven or in hell. Again, equal and opposite!

In yet another scenario, it is not just two opposing forces, but a series of natural phenomena that keeps things going in a cycle to maintain balance on earth. The clouds form and concentrate to give rain, which falls to the earth to water the fields and help crops grow. Then the water is evaporated by heat from the sun and goes back up into the sky, form clouds and come down again as rain; and the cycle continues.

The rotation of the earth around the sun gives us the day and the night, allowing us to wake and work and provide for ourselves; then sleep and rest and recover our energy to go again. It puts us through a cycle of gain, lose, gain and lose again; and this fuels our appetite for materials and comfort, creating in us a sense of livelihood. That is God's way of keeping everything interesting.

And the revolution of the earth around the sun results in the seasons of the year. We start at one season, ends up in another, then another and yet another. We plant, we reap, we feast and merry, and deplete our stock and need to replenish. We gladly go through the cycle over and over and over. We are fully engaged and never have time to get bored and languish. Our God is good and caring and made everything around us for our survival and sustainability, and even prosperity.

There is order and disorder, viability and decay, life and death. A seed that is sown into the soil decays for a new life to sprout. What was once alive (leaves and organic matters) dies and decomposes and becomes nourishment for new life (plant) that springs up from a dead seed. There is order and disorder, and even the disorder is ordered. That's the power of God as only He can.

There is usefulness and uselessness, enthalpies and entropies. Enthalpy is a partial measure of the internal energy of a system. Partial measure, because enthalpy cannot be directly measured. Only a change in it can be measured. In other words, it is the amount of energy in a system <u>capable of doing mechanical work</u>. This is useful energy; so termed, because it is available for man's use. And entropy, on the other hand, is a quantitative measure of the amount of thermal energy in a system that is <u>not available to do work</u>, in other words, a measure of disorder in a system.

The understanding of the natural things around us, on earth and in the universe as a whole, is definitely part of God's plan for mankind. Our understanding of nature further enhances our understanding of God. Pope John Paul II was quoted as saying: *"Science can purify religion from error and superstition. Religion can purify science from idolatry and false absolutes."*

I agree with him that science is not opposed to the understanding of God and obedience to God. Science enhances our appreciation of everything God is to us, and it shores up our confidence in His facts as written in the Bible. But I caution anyone who embraces science to be careful not to be misled by erroneous scientific pursuits and deceptive conclusions. They cause more damage to faith than anything else in the world.

Science, with extrapolations and assumptions which are contrary to the Bible, is lethal to faith and consequential to our livelihood and our eternal destiny. In the pursuit of knowledge, we have to ask God for wisdom so that we can correctly discern information and gain the right understanding. Knowledge discerned with wisdom leads to understanding that illuminates and enhances our experience with God.

That is why Jesus Christ said: *"I have come into the world as a light, so that no one who believes in me should stay in darkness." (John 12:46).* And you do not have to be highly educated to be illuminated as Christ declared in the passage above. He demonstrated that when He was in the world by gathering a bunch of regular people and filling them with unsurpassed knowledge that made the highly learned leaders of the Israelites look like a bunch of illiterates. By simply yielding your will to His Will, He opens up your heart to know Him.

There is good and evil; and there are many other equal and opposite forces in our lives. God is good and represents everything that is good and pleasing, and satisfying to the soul. On the other hand, he allowed Satan to rule the world with all the opposites. At every point there is God's quality, there is a detestable quality to match from Satan.

So, where there is good, Satan brings evil; where there is right, wrong; where there is joy, sorrow; where there is laughter, weeping; where there is happiness, anger; where there is strength, weakness; and so on.

Satan is simply a creation of God; just like everything else in the universe. But so that God can make His qualities known to and appreciated by the very creation (humans) He intends to share His heavenly kingdom with, He made Satan to be the backdrop on which He demonstrates His powers. We know this from His dealings with the Pharaoh of Egypt who refused to let the Israelites go.

In the Bible, God said to Pharaoh: *"For by now I could have stretched out my hand and struck you and your people with a plague that would have wiped you off the earth. But I have raised you up for this very purpose, that I might show you my power and that my name might be proclaimed in all the earth." (Exodus 9:15-16).* Pharaoh was to the Israelites what Satan is to the whole world.

There is nothing in the universe that compares to God, and as such, there is no power that equals Him. He made Satan a powerful nemesis, by allowing him powers that are much greater than that which any human being will ever possess. And because Satan is a spirit, he even impersonates God, boldly enticing and luring the greedy away from their Maker.

God inhabits everything He created so anything He extends to any creation still belongs to Him, and is controlled by Him. He can raise the power of the devil in a heartbeat to levels unsurpassed. For those who love Him and obey Him, He rises higher to overcome the devil and redeem them from his hands. To those who disobey Him and live wickedly, He allows them to experience the destructive powers of Satan so others will learn from their experiences.

Evil exists because good exists. God created Satan so God could demonstrate His powers against Satan. And without evil to contrast with, it will be difficult to understand good; and almost impossible to understand how difficult it is to do good. God allowed evil around us so we can develop aversion to it, and continuously choose good for our own goodness. You have not loved until you show love in the middle of injustice and unfairness.

Everything God created in the natural has a spirit God matches with it, that controls it. Every discrete particle; every living cell; every microorganism or virus; every discrete body tissue or body organ; every structure; every motion; every human emotion; every discrete object or system of objects; every celestial body or orbital system; every human establishment or human society—if it is real to man or exists in man's intellect—has a designated spirit in charge of it.

There is nothing in existence anywhere in the world or in the universe that does not have a controlling spirit assigned to it by God Almighty. God's entire creation is controlled by the Spirit of God. All the designated controlling spirits are integral parts of the Spirit that is God who is at work continuously throughout the world and the universe. He is an all-encompassing Spirit and leaves no stone unturned.

And because the Spirit of God fills the earth, the universe and beyond completely *(Ephesians 4:6-10)*; to God, moving anything from one point on the earth and the universe to another point on the earth and the universe, is like moving something from your left hand to your right hand. That is why everything God does within the earth and the universe takes only an instant.

Moving from the natural realm into the supernatural realms also takes an instant for the same reason. All of the spirits that exist everywhere are parts of the same Spirit that is God. Here is a passage from the Bible about all the spirits being integral parts of the Spirit of God:

*"**I am the LORD, and there is no other,**
    apart from me there is no God.
I will strengthen you,
    though you have not acknowledged me,
⁶ so that from the rising of the sun
    to the place of its setting
people may know there is none besides me.
    I am the LORD, and there is no other.
⁷ I form the light and create darkness,
    I bring prosperity and create disaster;
    I, the LORD, do all these things."* (Isaiah 45:5-7)

*"All your children will be taught by the LORD,
    and great will be their peace.*

[14] *In righteousness you will be established:*
Tyranny will be far from you;
  you will have nothing to fear.
Terror will be far removed;
  it will not come near you.
[15] *If anyone does attack you, it will not be my doing;*
  whoever attacks you will surrender to you.

[16] *"See,* **it is I who created the blacksmith**
  who fans the coals into flame
  and forges a weapon fit for its work.
And it is I who have created the destroyer to wreak havoc;
[17]   *no weapon forged against you will prevail,*
  and you will refute every tongue that accuses you.
This is the heritage of the servants of the LORD,
  and this is their vindication from me,"
declares the LORD." (Isaiah 54:13-17)

It is not cumbersome—nor does it take much of any time—for God to move anything from one part of Him, to another part of Him; just as it is not cumbersome—nor does it take much of any time—for a man to move anything from his left hand to his right hand.

The Spirit of God is God. And since the Spirit of God fills the earth and the universe completely, God fills the earth and the universe completely. And His Son Jesus Christ fills the earth and the universe completely. God is God the Father, God the Son and God the Holy Spirit. God the Trinity work is the One God and the creator of the earth and the universe and everything in them.

That is why nothing escapes God: He is over all of His creation; through all His creation and in all His creation. *(Ephesians 4:6)*

The spirit God puts in mankind is inseparably bound to the soul that is the human being. The Bible says that life is in the blood. That places both the soul and the spirit of man in the blood. And the Bible also says that the word of God is so sharp it divides the spirit from the soul. Here is that passage from the Bible:

"For the word of God is alive and active. Sharper than any double-edged sword, it penetrates even to dividing soul and spirit, joints and marrow; it judges the thoughts and attitudes of the heart. [13] Nothing in all creation is hidden from God's sight. Everything is uncovered and laid bare before the eyes of him to whom we must give account." (Hebrews 4:12-13)

God organized His world such that the natural is an extension of the supernatural. Since everything in the natural is controlled by a spirit, all the control for everything in the world is supernatural. The natural simply conforms to what the supernatural desires.

Good is real and evil is real. I would not call them particles because I do not understand their forms of existence, because that part was not given to me. But "good" has real existence, and "evil" had real existence, just the same way atoms, biological cells and light have real existence. But for the purpose of our discussion, I will call them the "good wave" or simple "good"; and the "evil wave" or simply "evil".

Our entire world is filled with these good and evil waves, just like our world is filled with the gases in our atmosphere (Nitrogen, Oxygen and Carbon Dioxide). And just as the Nitrogen gas in our atmosphere is inert (that is, non-reactive with the things in our world), the evil waves in our world is inert to the human person.

But unlike the Nitrogen gas, the evil gas becomes

reactive once it finds its way into a person—much in the same way a virus—t usually lifeless outside a host but becomes active once it gets inside a suitable biological host. This is God's grand design. And this is science at the highest level. The virus I mentioned above is actually one of God's agents of wrath; and so are the other disease agents.

God designed humanity to live exclusively on the good waves, generously given by the Spirit of God. The good wave was the only thing that affected the lives of Adam and Eve before they rebelled against God at the Garden of Eden. Their rebellion against God, then, unleashed the evil wave into our world, creating pollution in the world for the first time. The good wave then becomes saturated with the evil wave. In other words, every sample of the good wave had the evil wave in its mist.

Everything in our world became contaminated with the evil wave. That is why God said to Adam:

*"**Cursed is the ground because of you;**
   through painful toil you will eat food from it
   all the days of your life.*
*18 It will produce thorns and thistles for you,
   and you will eat the plants of the field.*
*19 By the sweat of your brow
   you will eat your food
until you return to the ground,
   since from it you were taken;
for dust you are
   and to dust you will return." (Genesis 1:17-19)*

This declaration from God to Adam that *"Cursed is the ground because of you,"* is literal. Everything is the whole world became contaminated from the evil wave that was unleashed when Adam and Eve sinned against God. It is this contamination that God was telling Adam about. And the contamination was not just on the ground, it was on everything in the world. God used the ground because everything in the world came from the ground.

The whole world had to adapt to a new destiny because of the emergence of the evil wave. The evil wave is the sole agent of death. It is the evil wave that started the once immortal bodies of Adam and Eve to become biodegradable, triggering the time clock of death in the human being. Sin is evil because sin generates and releases the evil wave into the world.

And what is sin? Here is what the Bible says is sin: *"**All wrongdoing is sin**, and there is sin that does not lead to death." (1 John 5:17)* In essence, sin is the commitment of evil. Disobedience of God is wrongdoing and as such sin. Adam and Eve disobeyed God and their disobedience changed the whole world forever—evil wave came into the world for the first time.

And the evil waves do not dissipate by itself. Nor is it consumed in any process on the earth. Evil wave builds up. And its detrimental effect on all lives on the earth intensifies due to the increased accumulation. Each additional sin committed anywhere in the world adds to the level of the evil wave in the world.

Sins concentrate evil waves around the people committing the sins. That is why the person who commits a sin is the one most affected by the evil wave his sin generates. Any human community that has more people committing moral atrocities has proportionately higher concentration of the evil wave within the community than a community that has fewer offenders.

This is what God was alluding to when He talked about the sins of the Amorites reaching its full measure. Their sins outputs evil waves that continued to build until it reached its full measure.

God sets the full measure of the evil wave any individual person or community could add to the surroundings. Here is the passage from the Bible:

"As the sun was setting, Abram fell into a deep sleep, and a thick and dreadful darkness came over him. [13] Then the LORD said to him, "Know for certain that for four hundred years your descendants will be strangers in a country not their own and that they will be enslaved and mistreated there. [14] But I will punish the nation they serve as slaves, and afterward they will come out with great possessions. [15] You, however, will go to your ancestors in peace and be buried at a good old age. [16] In the fourth generation your descendants will come back here, for the sin of the Amorites has not yet reached its full measure."" *(Genesis 15:12-18)*

And what constitutes full measure for one individual is different from what constitutes full measure for another. And the same is true for communities. The grace of God reduces the concentration of the evil wave within any community to lower the overall level for that community. Same is also true for an individual who had received the grace of God.

To understand what we are talking about in practical terms, visualize specialized sensors that operate 24/7, strategically built into each human being. These sensors only detect sin (wrongdoing). And each time it detects sin, it triggers activators within the individual.

These activators activate the evil waves within the person's immediate vicinity—we are continually surrounded by the good and the evil waves, just as we are continually surrounded by the atmospheric gases. And once these activators inside the human person are turned on by the sin sensors, they create access on the individual for the activated evil waves to enter the individual; and do havoc within the individual.

Remember, activated evil wave destroys the human life and

rots out the human body, and ultimately leads to death. Sin makes the evil wave that is ordinarily inert, active and dangerous for life!

This is what Apostle Paul was alluding to when he said to the

"In the following directives I have no praise for you, for your meetings do more harm than good. [18] In the first place, I hear that when you come together as a church, there are divisions among you, and to some extent I believe it. [19] No doubt there have to be differences among you to show which of you have God's approval. [20] So then, when you come together, it is not the Lord's Supper you eat, [21] for when you are eating, some of you go ahead with your own private suppers. As a result, one person remains hungry and another gets drunk. [22] Don't you have homes to eat and drink in? Or do you despise the church of God by humiliating those who have nothing? What shall I say to you? Shall I praise you? Certainly not in this matter!

[23] For I received from the Lord what I also passed on to you: The Lord Jesus, on the night he was betrayed, took bread, [24] and when he had given thanks, he broke it and said,"This is my body, which is for you; do this in remembrance of me." [25] In the same way, after supper he took the cup, saying, "This cup is the new covenant in my blood; do this, whenever you drink it, in remembrance of me." [26] For whenever you eat this bread and drink this cup, you proclaim the Lord's death until he comes.

[27] So then, whoever eats the bread or drinks the cup of the Lord in an unworthy manner will be guilty of sinning against the body and blood of the Lord. [28] <u>**Everyone ought to examine themselves before they eat of the bread and drink from the cup.**</u> *[29]* <u>**For those who eat and drink without discerning the body of Christ eat and drink judgment on**</u>

*themselves. [30] **That is why many among you are weak and sick, and a number of you have fallen asleep.** [31] But if we were more discerning with regard to ourselves, we would not come under such judgment. [32] Nevertheless, when we are judged in this way by the Lord, we are being disciplined so that we will not be finally condemned with the world.*

[33] So then, my brothers and sisters, when you gather to eat, you should all eat together." (I Corinthians 11:17-33)

The evil wave activated by our sins makes us "weak and sick" and even worse; it kills us. So, it is not God Who brings sufferings and all the distasteful things into our lives. It is us that do it to ourselves, by choosing not to obey God but instead, follow the directions of our own minds.

When you are emitting the evil wave, people around you can pick it up. Their feelers (sensors for the evil wave) pick it up; in much the same manner their feelers pick up the preponderance of good wave when it is in their vicinity. When people say, "I am getting a bad vibe from this guy," what they are really saying is that they are sensing the evil wave that is coming from this individual and it repels them.

People can sense the good wave that is coming from someone. People can also sense the evil wave that is coming from someone, unless they are into the same thing that the person emitting the evil wave is into. In that case, it becomes a high for them because the evil wave they are sensing is increasing the intensity of the one already in them.

That is why a mob works in concert in its destructive activities. The Evil wave that is being generated from different sources within the mob quickly intensifies the effect of the wave on members of the mob and continues to fuel their evil deeds until it climaxes in them; and they wrought their destruction.

Out of His great love for mankind, God had designed all human beings with a buffer against the evil wave, because He knew the evil wave would inevitably come into the world through man's disobedience of Him. You can be surrounded by, and submersed in, evil wave and not be affected by it at all— if you have not destroyed the natural buffer against it that you are born with. And it takes a lot of deliberate rebellion on your part to break down this natural buffer you are created with.

That is why it is unnerving to get into rebellion for the first time. You wonder what would happen and whether you will still know what to do with yourself after you have rebelled. You are hesitant but something continues to push you. You take a little baby step at a time until you finally break through that buffer and let evil in.

And once you get there, you realize that there is not going back. At that point you decide to deal with whatever the consequences are and move on with your life. And you would definitely have people who rally on your side to help you minimize your guilt, even calling you a hero for summoning the courage to come to the other side.

Sin is the multiplication of the evil wave. Each sin committed by mankind increases the amount of the evil wave in the space that we all share and live in. It is not just the atmospheric space but also the ground and the sea. Only the grace of God has the power to degrade and destroy the evil wave.

Sexual sins are the biggest contributors of the evil wave in the world societies. They are so prevalent and most in the communities where they happen have no knowledge of what is happening. And because sexual sins are mostly committed in hiding, the perpetrators of sexual sins believe that since no one sees their deeds, nobody gets hurt. That is farthest from the truth. Every single human sin contributes new evil waves to what is already in the society. And this is without failing.

You cannot mix an acid and a base and expect there would not be a reaction, and the generation of outcomes. Similarly, you cannot commit sin and expect evil waves not to be generated and added to what is already in existence within the society. It is automatic and has been right from the original human rebellion against God at the Garden of Eden. Outcomes always happen in life's circumstances. Outcomes always happen in spiritual circumstances! The natural is an extension of the spiritual. This is God's design.

When conditions are suitable for something in God's world and universe to happen, that thing invariably happens. That is why it is laughable when human beings set up the conditions for things to happen as God designed them to happen, and the human beings take credit for what had happened; and aspire to replace God because of it. God had just cracked the window a little and gave us a peak into His mystery. We are not going to take over God's world, because we cannot escape God's grip on us and our world.

Every time sin occurs, it releases evil waves, and the evil waves add to the world's existing concentration. The world is oblivious of this, but our ignorance would never change the outcome of our bad deeds. They add to the level of the destructive evil wave that is tarnishing everything in and around our lives.

Grace is a spirit; the good wave is a spirit and the evil wave is also a spirit. Through the grace of God, the good wave overcomes the evil wave. And as grace increases, the evil wave dissipates. That is why a believer in Jesus Christ gets the Holy Spirit of God to dwell inside him or her.

As you grow in godliness, the Holy Spirit that is in you

intensifies the good spirit (good wave) that is in you to levels that matches your godliness—not your religiosity. And when a difficult situation arises in your life in which God decides to intervene, His grace intensifies the good wave in you to enable you to overcome the difficult situation.

For the atmospheric gases—Nitrogen, Oxygen and Carbon Dioxide—plant's intake of carbon dioxide and output of Oxygen in photosynthesis; and animal intake of Oxygen and output of carbon dioxide in respiration, help to maintain the concentrations of these gases in the atmosphere.

But the concentrations of the evil and the good waves do not remain constant in any human society. They fluctuate based on the morality in the community. A community that does a lot of good, and much less evil, has a proportionately higher grace than a community that does a lot of evil, and little good. The level of the evil wave within a society could build to suffocating levels. That is when the wrath of God descends on the community.

This is not to say that a community that is more prone to bad natural disasters has greater propensity for evil than a community that has less occurrences of disasters. However, distasteful events happening at different parts of a nation at higher frequency than normal, signals God's dissatisfaction with the nation, not necessarily the parts that are directly hit by the disasters.

World governments and world communities frequently embark on projects geared at reducing pollution in our communities. Yet, humanity has failed to believe God and obey His commands to avoid evil because of the terrible effects the evil wave has on the human life. The main reason for this is that humanity did not recognize that evil is real and spews poison into our world every single time it is committed.

In just the same way methane and other toxic gases are generated and released into our atmosphere from toxic wastes and human pollutants, the evil wave is generated and released into our surroundings every time someone commits a sin, knowingly and deliberately or unknowingly and non-deliberately.

The simple graph below represents these good and evil waves and their effects on human beings.

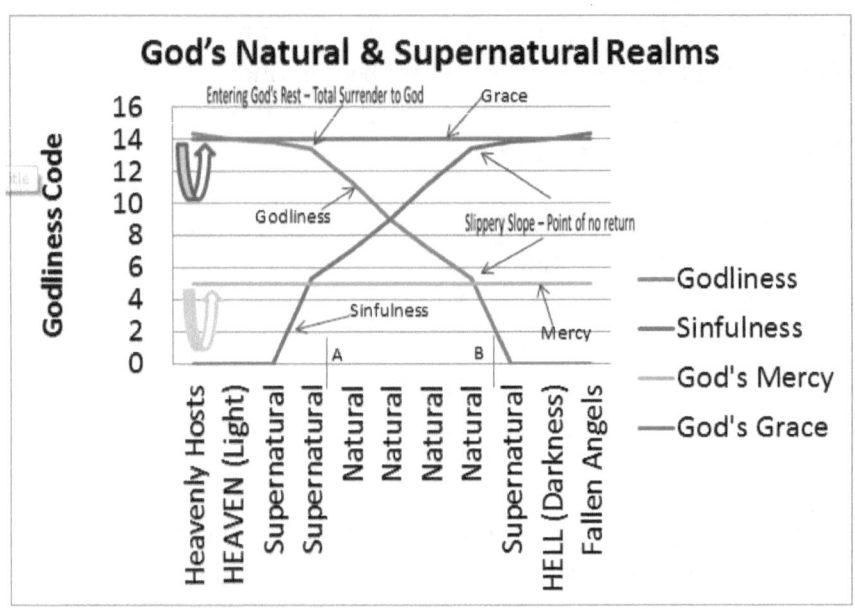

Godliness is the only antidote to the evil wave, because godliness attracts the grace of God. Godliness is the direct opposite of sinfulness. As you increase in godliness, your propensity to sin decreases. Many of us are content with simply having a balance between how much good we do and how much evil we allow ourselves to get into.

But as you can see from the graph above, that is only the start of your journey in godliness—unless you are climbing out from a moral pit. It is not a place you want to spend the rest of your life at.

You can see that godliness is not a religious thing. Godliness is the correct path to life. Jesus Christ encouraged religion because getting together as a group helps members of the group support one another in their journey to godliness; and also offer them the fertile ground to demonstrate their obedience to God. Godliness is life itself!

Every human being must aspire to continue to grow in godliness day after day. We must all aspire to enter God's rest. Once you enter God's rest, the slope becomes more gradual; sin would has virtually very minimal attraction to you; and the effect of the evil wave on you is also minimal, if not non-existent. God is satisfied with you and your life continues to be a light that shines in the darkness; like that of Jesus Christ when He was in the world.

As grace increases, it protects the godly person from the decaying effects of the evil wave. The evil that you commit does not affect only your own life. It affects the lives of others around you, especially those who share something in common with you in their lives: like your spouse, your children, your parents, your siblings and your friends, to name a few.

That is why God calls all humanity to godliness and love. Godliness creates a spiritual buffer around the godly person. And as their godliness increases, the size of this buffer expands to eventually surround all of their loved ones and all their belongings—in the same way God did it for Job—*"**Have you not put a hedge around him and his household and everything he has?** You have blessed the work of his hands, so that his flocks and herds are spread throughout the land." (Job 1:10)*

The line that represents God's grace in the chart above is not a fixed line; it rises as God desires for every given circumstance or a person. Grace only comes from God as a gift to help humanity combat the desire for sin and triumph over it. Grace is a toll God gives to mankind to use actively; and not to allow to stay dormant. Grace is given so that mankind can do more good.

There is also a line on the chart that represents God's mercy, and that line is placed lower than the line representing God's grace. God's mercy lifts you out of the strangling effects of the evil wave, and God's grace restores you and pushes you up. Therefore, mercy and grace work together in every given situation of human redemption.

The line that represents mercy on the chart could deep as low as God desires it to dip, to rescue a soul that is falling into the pit of despair. God's mercy could go all the way to the bottom of the pit and rescue a soul that is already drowned. And then God's grace takes over, rehabilitates the soul and restores the soul to its former glory. And grace does not stop there. Grace could lift the soul much higher than the soul's original position in life.

Mercy can tag along with grace for a while too. When anyone who was rescued by God's mercy from the devastation of the evil wave struggles to remain on the right path to God's righteousness; God's mercy continues to pick them up and steady them on their feet as they falter; and God's grace continues to do the restoration, reinstatement and uplifting of the soul. This collaborative effort continues until the soul could handle the journey with just grace.

The points "A" and "B" on the chart represents the thin veils that separate the natural from the supernatural on both sides of the natural—the heavenly side and the dark side.

You can see from the chart that the natural straddles between the heavenly supernatural and the dark side. Darkness has nothing to do with the light, therefore God sandwiches the natural between both the good side and the evil side. In other words, spirits from either side operate in our world, pulling us to their respective sides.

This representation on the godliness chart above is similar to the light spectrum chart, where the visible light (which we see) is sandwiched between two unseen extensions of light—bordered at the far ends by the gamma ray on the one side and the radio wave on the other.

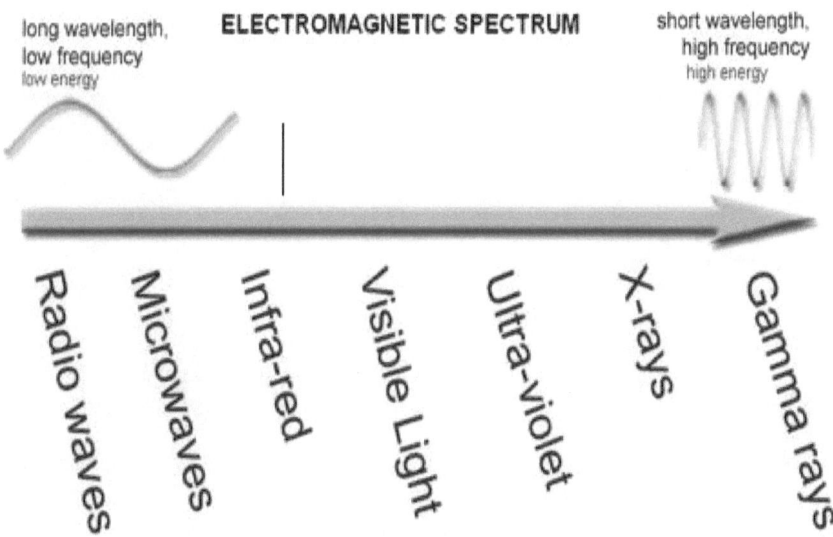

The radio wave side represents light with long wavelengths, low frequency and low ionization energies. And the Gamma ray side represents light of short wavelengths, high frequencies and high ionization energies. The high ionization energy (the gamma ray side on the chart) represents power and strength and operates in clearly defined ways and acts decisively, obliterating everything evil on its path.

The low ionization energy (the radio side on the chart) represents the dark spirits with their pretentiousness; always operating under disguises; always lurking behind shadows; and always seeking to inflect damage to humankind through deceit.

And the visible light which had been known to man, from mankind's first day on the earth, sits right in the middle just like the natural sits between the good supernatural and the evil supernatural on the godliness chart.

Chapter 5

GOD NEVER CHANGES

Just as Eve was never fully convinced that what God said would happen would, indeed, happen; many Christians doubt the plausibility, or applicability of one thing or another that is said in the Bible. It was Eve's unbelief that caused her to fail, not the gullibility of the serpent. That was simply the very small tap needed to push her overboard. When you have unbelief, you are ready to fall over.

God can never be outdated, because He is the only one who is doing the dating. Everybody else is affected by His dating. And that is why we all live, flourish for a while and, then, we die and our persons and personalities become completely extinct, while God lives on. When people talk about modernizing the things of God, they are indirectly suggesting that God's wisdom is outdated and needs to be upgraded.

That is why many Christians have issues with a lot of the events of the Old Testament, or many of what Jesus Christ and His apostles commanded in the New Testament. Most of the things believers agitate to see changed in the Christian churches are actually warned against in the Bible, yet many do not care about the warnings in the Bible. They simply want to update these things to 'catch up with the times.'

If God created time and space, the universe and

everything in it, what makes anyone think that time would outrun God? How can something He created, and knows how it began, and determined how it will end, somehow outrun Him; such that He now needs to be upgraded to keep pace with it? That kind of thinking only demonstrates how little we think God is.

God is above and beyond everything He created, or He would not be able to create, control, regulate and maintain them. You hear arguments about how Jesus came and changed the Law. Jesus Christ is one part of the Trinity that is God, so His coming to the earth would not negate God, for then He will be contradicting Himself.

What we are doing in our agitation is demonstrating our unbelief. That is why believing in Jesus Christ is not simply by saying that we do, but by earnestly seeking to know Him and understand Him so we can partake of His salvation, which has been given to all freely but can only be received through faith. And faith is: *"being sure of what we hope for and certain of what we do not see."* *(Hebrews 11:1).*

And we can only be sure and certain of what the Bible says if we put all our trust on Jesus Christ — and not in our own natural power of logic and worldly experiences. Unbelief is living without faith. The law of God is applicable to all humanity and equally valid at all ages of human existence. The God of the ancients is the same God of the modern days, and he never misses a beat. Without the Old Testament, the New Testament has no meaning. And without acknowledging the New Testament, a believer gets stuck in time past, while the rest of God's people move on to seize the prize.

That is why Jesus said *"Do not think that I have come to abolish the Law or the Prophets; I have not come to abolish them but to fulfill them. I tell you the truth, until heaven and earth disappears, not the smallest letter, not the least stroke of pen, will by any means disappear from the Law until everything is accomplished."* *(Matthew 5:17-18)*. With this statement, He affirmed that everything contained in the Law of Moses is still valid beyond measure. It is all absorbed into the new covenant.

They are as valid now as they were valid then. And in the very next verse, He added: *"<u>Anyone who breaks one of the least of these commandments and teaches others to do the same</u> will be called least in the kingdom of heaven, but whoever practices and teaches these commands will be called great in the kingdom of heaven. For I tell you that unless your righteousness surpasses that of the Pharisees and the teachers of the Law, you <u>will certainly not enter the kingdom of heaven.</u>"* *(Matthew 5:19-20)*.

To understand what Jesus Christ meant by having a righteousness that surpasses the righteousness of the Pharisees and the teachers of the Law, consider the following passage: *"Woe to you experts in the Law, because you have taken away the key to knowledge. You yourselves have not entered, and you have hindered those who are entering."* *(Luke 11:52)*.

In the same Scripture in the gospel of Luke, Jesus Christ accused the Pharisees of cleaning the outside to impress the people and earn their respect, when their inside is full of greed, deceit and wickedness. He condemned their practice of appearing to be living holy lives when, in reality, they were all pretenders.

So, the God of the Old Testament is the same God of the New Testament and will remain the same forever. God does not change, and does not have any need to because He gets everything right the first time. Listen to what the Bible says about that: *"I know that everything God does will endure forever, nothing can be added to it and nothing taken from it. God does it so that men will revere him."* (Ecclesiastes 3:14).

If God said it, He said all that need to be said — no more and no less. If you want to know more, He will teach you but you have to learn in humility and reverence to Him. He is still working today, taking charge of His universe, keeping everything in his view so that only His Will may be done, because only He has the perfect knowledge on how best anything works.

And to convince us all of the futility of second guessing God, the Bible went further and says: *"I have seen the burden God has laid on men. He has made everything beautiful in its time. He has also set eternity in the hearts of men; yet they cannot fathom what God has done from beginning to end."* (Ecclesiastes 3:10-11).

We are the ones that need to change by striving to have the mind of Jesus Christ, as the Bible tells us. Our wisdom is foolishness to God, even with the collective advancement we have made in the world. If we fail to see things the way God sees them, it is our loss, not His. And we are the ones who will pay the price by earning places in the lake of eternal fire for our disobedience.

Look at this passage in the gospel of Matthew: *"Not everyone who says to me, 'Lord, Lord,' will enter the kingdom of heaven, but only he who does the will of my Father who is in heaven. Many will say to me on that day, 'Lord, Lord, did we not prophesy in your name, and in your name drive out demons and perform many miracles?' Then I will tell them plainly, 'I never knew you. Away from me, you evil doers!'"* (Matthew 7:22-23).

Jesus is unequivocally speaking about believers in

this passage, and not unbelievers. He is talking about people who believed they have received Him as their Lord and Savior but really haven't. No unbelievers will be prophesying in the name of Jesus Christ. Nor would an unbeliever be driving out demons or performing miracles in the name of Jesus Christ, either. Anybody who uses His name for anything believes that He has received the authority to do so.

Yet, Jesus Christ is unmistakably saying here that, there will be people who think they are doing the right thing in His name, when they, in reality, are serving their own purposes or those of others; and not doing anything to obey Christ's commands and be saved. Jesus is also saying that people like that will be denied entry into the kingdom of heaven, and He would simply say to them, 'I never knew you. Away from me you evil doer!' That sounds harsh, but that is the only reason why Jesus Christ made it this clear: to prevent anybody from being deceived into doing the wrong thing.

This is exactly why the Bible warned us: *"Your attitude should be the same as that of Christ: Who, being in very nature God, did not consider equality with God something to be grasped, but made himself nothing, ...Therefore my dear friends ... continue to work out your salvation with fear and trembling, for it is God who works in you to will and to act according to his good purpose."* *(Philippians 2:5-13).*

Having an attitude like Christ means staying humble and working towards one's salvation with humility and dedication; without thinking more highly of oneself than he really is. And this is why the Apostle Paul added a few verses further down, *"Not that I have already obtained all this, or have already been made perfect, but I press on to take hold of that for which Christ Jesus took hold of me. Brothers, I do not consider myself yet to have taken hold of it. But one thing I do: Forgetting what is behind and straining toward what is ahead, I press on towards the goal to win the prize for which God has called me heavenward in Christ Jesus.*

"All of us who are mature should take such a view of things. And if on some points you think differently, that too God will make clear to you. Only let us live up to what we have already attained. Join with others in following my examples, brothers, and take note of those who live according to the pattern we gave you. For, as I have often told you before and now say again even with tears, many live as enemies of the cross of Christ. Their destiny is destruction, their god is their stomach, and their glory is in their shame. Their mind is on earthly things." (Philippians 3:12-19).

The underlined part of this passage clearly indicates that Apostle Paul was not talking about unbelievers, rather, he was talking about believers of Jesus Christ, who, instead of focusing on gaining the approval of Jesus Christ in their lives and their ministries; are rather allowing their desires for fame and fortune to derail them. That is why he said in the passage 'and now say again even with tears'. Those people have failed to continue on the path of salvation, and instead, went after earthly rewards, mistakenly tagging them as rewards from God for their 'good stewardship'. But in reality, they are total failures with regard to the kingdom of heaven, which is what they signed on to in the beginning.

And to be sure that everybody understands what salvation entails; the apostle uses himself as an example. Not even he has attained the holiness that is needed for anyone to see God. But to be sure that he does when that time comes, he does not look behind, but continue to strain forwards towards the prize, almost single-mindedly, because the prize is that important. He was not going to let anything deny him that chance, especially not self-aggrandizement.

And he is advising all of us who are mature to be of such mind, and if we think differently on some points, to look to God to make them clear to us; instead of making wrong assumptions on our own and getting sidetracked. But, are we really taking his advice today? It appears we have better edge on things than his advice could help us with. That is sad, but there will be consequences.

The God of Science

Chapter 6

EVERYTHING IS WHATEVER JESUS CHRIST SAYS IT IS

Everything is whatever God says it is, and He does not call anything something unless it is, indeed. He created everything in the universe out of nothing and calls them something. And that is what we know them to be because that is what He said they are. The sun is the sun because, God said it is the sun. So is the moon, the earth, the stars, the seas and everything else we know today.

He said He created the universe out of nothing and we believed Him completely. *"By faith we understand that the universe was formed at God's command, so that what is seen was not made out of what was visible." (Hebrews 11:3).* Science has also confirmed that the universe was created out of nothing. God said that He made us in His own image and we believed Him.

He told us that He hung the earth on nothing (in the Book of Job) and we believed Him — that was four thousand years before any scientists finally confirmed that the earth was suspended on nothing. We believed simply because He told us so. He told us that a single man, Noah, obeyed Him and for 120 years built a boat large enough to provide refuge to his family of eight, two each of all the different species of animal in existence in the whole world, and one year's food supply for his family and all the animals with them in the boat; and we believed Him.

We did not wait for the evidence that science uncovered before we believed God. We believed Him because He wanted us to have faith in Him and accept whatever He says to us as factual without waiting for verification by the world. That is what faith is all about: *"Now faith is being sure of what we hope for and certain of what we do not see." (Hebrew 11:1).* And once you develop that faith in Him, it never leaves you. Doubting any of what He says in the Bible is not even remotely a consideration for you.

Many who went before us, and some even now, have the tendency to pick and choose what to believe in the Bible and what not to believe, sometimes under the guise that what is written takes a deeper philosophical knowledge to understand. That is not the kind of faith the Christians are called to. Whatever the Bible says, it means literarily, and we are called to believe everything in the Bible by faith. That is why the Bible says, *"And without faith it is impossible to please God, because anyone who comes to Him must believe that he exists and that he rewards those who earnestly seek him." (Hebrews 11:6).*

The Bible was not written for the sophisticated minds only. It is written for anybody who is thirsty for the wisdom and direction of God, because the Holy Spirit of God goes through the Scripture with you in your search for the truth of God, if you are earnestly seeking God. Anyone who has good enough understanding of language can study the Bible with good understanding, especially in conjunction with listening to Bible teachers and preachers of the gospel of Jesus Christ.

I am not talking about the things we have difficulty understanding in the Bible, because the Holy Spirit does not reveal everything in the Scripture to you at once. What you do not understand yet, continue to search and He will eventually reveal it to you. He knows your needs and will never deny you anything that you need to be able to live a holy life, for that is the Will of God for your life. I am talking about things we read in the Bible but do not see how they can be, because we are measuring the word of God in terms of our limited earthly experiences and understanding.

When Jesus Christ asks us to break bread and wine in remembrance of Him, for the bread is His body and the wine his blood; He did not intend that to be symbolic. It is real. Thinking of it as symbolic is the same as thinking of telling His apostles to wash one another's feet as being symbolic. If washing the apostles' feet was merely a symbol, why did Jesus tell Peter, *"Unless I wash you, you have no part with me."? (John 13:8)*. It is that serious.

Jesus Christ has no use for symbolisms. And nobody understands this better than the apostles, including Paul. Jesus Christ is the Savior of the world. He is the real deal and nothing about Him is symbolic. God gave Moses a lot of symbolic things, reserving the real things until Jesus Christ came. Once Jesus came, He fulfilled everything. He is the Life, the Light, the Truth and the Way. What else is there aside from these? He is everything that means anything.

Anything He is not simply does not exist. And if He is all there is, what use is symbolism to Him? So, if the redeemer of humanity Himself says that the bread we break in His name is His body, who are we to tell Him it is not his body but common bread? Or that what He said is His blood is not His blood but wine? If we do that, we are simply challenging His divine knowledge, and that constitutes unfaithfulness. Nobody is greater than His master.

If Jesus Christ intended the cup and the bread as symbols, He would have addressed them as such. But when He instructed His apostles, He told them how to treat what they were receiving because it was no symbolism — they are what He says they are. *"__This is my body__ which is for you; do this in remembrance of me."* And for the wine, He gave it to them and said *"__This cup is the new covenant in my blood__; do this, whenever you drink it, in remembrance of me."*

Just like Peter questioned the wisdom in Jesus washing his feet but promptly yielded to be washed when corrected, if any of the apostles have questioned why Christ was calling what was clearly bread and wine His Body and His blood, that apostle would have been corrected the same way and he would readily understand.

Jesus said to his apostles: *"You know the way to the place where I am going." (John 14:4).* To this, Apostle Thomas replied: *"Lord, we don't know where you are going, so how can we know the way?" (John 14:5).* And Jesus Christ answered: *"I am the way and the truth and the life. No one comes to the Father except through me." (John 14:6).* Problem solved! Because Jesus Christ told them that He is the way, they got it. Not only did that resolve their curiosity as to where exactly He was going, it settled their puzzlement as to how to get there. They simply believed Him.

Philip, then, requested from Jesus: *"Lord, show us the*

Father and that will be enough for us." (John 14:8). And Jesus answered, with a little surprise in his voice: *Don't you know me, Philip, even after I have been among you such a long time? Anyone who has seen me has seen the Father. How can you say, 'Show us the Father?' Don't you believe that I am in the Father, and that the Father is in me? The words I say to you are not just my own. Rather, it is the Father, living in me, who is doing his work. Believe me when I say that I am in the Father and the Father is in me; or at least believe on the evidence of the miracles themselves. ..." (John 14:9-14).* This, too, is settled. Because, He said so, the apostles believed. They dropped their request for evidence and trusted in His words.

Now, is Jesus Christ being the way the truth and the life a symbol or is it real? Is Jesus Christ being in the Father and the Father being in Him a symbol or is it real? Jesus Christ is everything He claims to be, and everything is what He says it is. That anybody has a hard time fathoming what he said does not make it a symbol or a figure of speech. See what He said in the following passage: **"Though I have been speaking figuratively, a time is coming when I will no longer use this kind of language <u>but will tell you plainly</u> about my Father. <u>In that day</u> you will ask in my Father's name...."** *(John 14:25-28).*

And immediately following this comment, His observed, *"<u>now you are speaking clearly and without figures of speech</u>. Now we can see that <u>you know all things and that you do not need to have anyone ask you questions</u>. This makes us believe that you came from God." (John 14:29-30).* And to this, Jesus replied: *"<u>You believe at last!</u>" (John 14:31).*

This is the time — 'in that day', the day Jesus Christ promised in the preceding passage to start speaking to us (His disciples) plainly and not figuratively. And He made us a promise that whatever He said to us that we did not understand the Holy Spirit of God will teach us all things and remind us of everything He had said. Here is the passage from the Bible: *"... All these I have spoken while still with you, But the Councilor, the Holy Spirit, whom the Father will send in my name, will teach you all things and will remind you of everything I have said to you."* *(John 14:24-26).* And on that day of the Pentecost, He fulfilled all of these, and the Holy Spirit is still with us today, teaching us all things.

Jesus Christ is talking to us plainly now. Figure of speech is designed for the unbelievers, but to us He speaks plainly. What we receive at the Holy Communion is His body and His blood, as He told us, and not whatever else anybody else is making them out to be.

That is why Apostle Paul said the following to the Corinthians: *"For anyone who eats and drinks without recognizing the body of the Lord eats and drinks judgment on himself. That is why many among you are weak and sick, and a number of you have fallen asleep."* *(1 Corinthians 11:29-30).* The bread and wine we share at Eucharist (or the Holy Communion) are no longer bread and wine. They are the body and the blood of Jesus Christ, as He said they are. And there is power in them, to heal and to restore us, when we partake of them as directed by Jesus Christ. He is the real deal and has no use for symbolism.

Ordinary bread and wine do not have the power Apostle Paul was alluding to in this passage from first Corinthians. Ordinary bread and wine will not cause anyone to become weak, sick or dead. We consume bread and wine to maintain our energy, and to liven up the body. The fact that the what we receive at the Holy

Communion work in reverse (when taken without recognizing them as what Jesus Christ said they are) testifies to them being what Christ said they are; not what an unbelieving mind sees them as.

In a Scripture in the Book of Exodus, Moses come up to the burning bush and was told by God to throw his staff down and Moses did. His staff which was made of wood turned into a snake, and Moses moved back in awe of what had just happened. The moment that stick became a snake, it exhibited all the characteristics of a snake because it had, indeed, become a snake. That is why Gold told Moses to pick it up by its tail. *(Exodus Chapters 3 & 4)*

If Moses had gone anywhere close to the head of the snake, he would have been startled more by the snake, because the snake will strike at him as snakes do when they are threatened. Because God is the one who created everything and gave them rules on how to behave and react, God would not deny the snake an opportunity to be a snake for that is what Moses' staff has become.

This was necessary also to convince Moses that what he was looking at was not magic, but a real snake. God told Moses to pick the snake up by its tail — not the midsection or the head. And when Moses did as he was told, the snake turned back into his staff, but not before he had it in his hand. If the wooden staff was not turned into a real snake, Aaron's staff would not have been able to swallow the other two snakes at the Pharaoh's palace. Its size must have apparently increased after it swallowed the other snakes, but when it turned back to Aaron's staff, it returned to its original shape and size; not bulging in the midsection because of the other snakes it ingested when it was a snake.

Moses was very familiar with his staff and knew it was a dead piece of wood. Yet before his very eyes, his dead piece of wood became a dangerous snake and he felt threatened by it, because he realized he was no longer looking at his staff but rather a deadly snake that could kill him. Moses believed what he saw, and knew it was real because He believed that God has the power to change anything, after all, He made the universe out of nothing.

Similarly, following God's instruction, just sticking his hand into his pocket caused Moses's hand to become leprous. Moses knew it was not symbolic, but real. If God did not reverse the leprosy by having him stick his hand once more into his pocket, the leprosy could have spread all over his body, because Moses' hand was, indeed, infected with the real disease. God was simply proving a point to Moses, which is that He, God, is indeed, the creator of heaven and earth. Moses was convinced he was struck with real leprosy, not a symbolic one.

So if God says that something is, it becomes. That is why the Scripture says that Abraham is *"our father in the sight of God, in whom he believed — the God who gives life to the dead and calls things that are not as though they were." (Romans 4:17)*. Simply, if God wants anything to be, He calls it that and it becomes. It is the same way with Jesus Christ, His Son. That is why the Scriptures say of Him, *"For God was pleased to have all his fullness dwell in him, and through him to reconcile to himself all things, whether things on earth or things in heaven, by making peace through his blood, shed on the cross." (Colossians 1:19-20)*.

God was pleased to have all His fullness in Jesus Christ. That is why Jesus Christ had the power to perform all the miracles He performed in His ministry. When in Cana, He asked the groom's servants to pour water into clay pots and take a sample to the wine taster; the water became the sweetest wine at the wedding ceremony, simply because Jesus Christ wanted the water to become wine.

Jesus did not touch the pots of water or pray over them. He simply commanded the servants to fill the clay pots with water, and after they were done, He simply commanded them to take a sample to the wine tester. It was ordinary water poured into the pot by others at His command. But because He is 'the God that calls things that are not as though they were,' the water became wine, without anybody adding anything to it to transform it.

He did not tell the wine tester or anyone else at the party to pretend that they were drinking wine when they, in reality, were drinking water. The wine He caused to come into existence was real, not magic or symbolism. The guests at that ceremony all ascertained for themselves that they were, indeed, drinking the best wine at the ceremony; because Jesus Christ had caused the water to become the best tasting wine at the ceremony. And He did not require anybody at the ceremony to accept the water as wine by faith, because it was not necessary. <u>The water had simply become wine, because Jesus Christ called it wine.</u>

It becomes imperative, therefore, for any Christian who questions anything that comes out of the mouth of Jesus Christ, to reassess his faith in Him. If Jesus says something is, it is, because He made things what they are; and has the power to change them into whatever He wanted to change them into.

If a piece of lifeless wood can turn into a snake and revert back to a piece of wood; and leprosy could instantly affect a hand and the hand, instantly cured; and water can turn into wine from the moment it got into a clay pot to the moment it was drawn out of the clay pot; why would bread not turn into Christ's precious body, and wine, into His precious blood? It requires faith, and that is what we are called to have or we have not achieved anything. And that is why believers are healed or restored when they partake of the bread and wine at the Holy Communion.

Our problem with certain things that are said in the Bible is that we always try to bring God and Christ and the Holy Spirit to our own level of understanding to convince ourselves that what we are asked to believe is sensible. If it does not fit our understanding, then there's got to be something else. It does not work that way. We need to be elevated by the Holy Spirit of God to the level of Christ to understand the endless possibilities within His power. He is everything He says He is and more. And whatever He calls anything is what it is, and there is no other way around it.

We all have to remember what the apostle Paul said: *"Now we see but a poor reflection as in a mirror; then we shall see face to face. Now I know in part; then I shall know fully, even as I am fully known."* *(1 Corinthians 13:12)*. Faith is all we need now if we are to be successful in our hope to enter the kingdom of heaven. Through our faith in Christ, we are fully known and our place in the heavenly kingdom is secured.

According to the passage, all we currently know about God is not all there is to know. That is also evidenced from other passages in the Bible, for instance, when Apostle Paul overheard conversations in paradise which *'no man was permitted to tell'* (2 Corinthians 12:3-4); and when Apostle John, in the Book of Revelation, was instructed by one angel not to write down something that was said to him about the end time events, because it would remain a mystery until the end time. *(Revelation 10:4)*

It makes sense, then, to conclude that the best textbook for the study of all human fields of professional specialties is the Bible; complimented by other books in the various fields of learning! Here is a sample:

Academic Discipline: Seismology, Geology and Earth Science
Recommended Textbook: The *Holy Bible*, plus ...

Academic Discipline: Infectious Diseases Treatment, Control and Prevention
Recommended Textbook: The *Holy Bible*, plus ...

Academic Discipline: Heart Disease, Mental Illness, Acute Confusion, Derision
Recommended Textbook: The *Holy Bible*, plus ...

Academic Discipline: Biology, Biochemistry, Microbiology, Botany
Recommended Textbook: The *Holy Bible*, plus ...

Academic Discipline: Physics, Physical Chemistry, Anatomy & Physiology
Recommended Textbook: The *Holy Bible*, plus ...

Academic Discipline: Philosophy, Psychology, Sociology,
Recommended Textbook: The *Holy Bible*, plus ...

Chapter 7

THE BIBLE IS SCIENCE BEFORE
MANKIND COINED THE WORD "SCIENCE"

True Science affirms God and the truth of the Bible! But the devil has managed to tack on a lot of lies and pretentiousness to true science in order to get science to fulfill the prophecies of the end time. Science does not alienate the Christians as much as the Christians alienate science!

There are mostly two classes of Christians: those who distrust science and so stay away from it; and those who do not know the distinction between true science and the pretentious one that is dubiously also dubbed science, and yet they quickly point to science as the evidence that the Bible is correct. Any Christian embracing science has to be able to make the right distinction or he will be swept away in the lies that are mixed in with true science.

And then there is a third category of Christians who understand that true science can only affirm the truth in the Bible and so are willing to approach the things of the Bible methodically as in science, instead of everyone saying whatever feels right to them. This should be the position of all Christians so that errors may be avoided. The Bible is the truth, and through scientific approach, the truth of the Bible can only be affirmed; not disproved, because the Holy Spirit of God, who gives discernment to anybody who faithfully and sincerely studies the Bible, never misleads genuine efforts.

The Bible is the origin of all the world's academic disciplines. The early scientists all started with the investigation of the claims made in the Bible about the earth and the universe. That is why the Catholic Church was at the center of all early scientific hypotheses and discoveries.

The church supported science because the church believed that since God is the truth, and God' word truthful; and science was designed to seek the truth—and nothing but the truth—that there is no way science could become something evil and averse to God.

The church was right in its assumption. However, science is something more mysterious than the church understood—science is the False Prophet and was prophesied to come up within the church **("the land")**. Science would eventually become an umbrella for anti-God hypotheses like evolution to draw humanity away from God.

While true scientific discoveries and advancements mesmerize the public, the unfounded and anti-God hypotheses like the so-called Theory of Evolution tacitly accomplish the will of the devil who started the aversion in the first place and promoted it through the centuries; through individuals who are more interested in personal recognition and rewards that they are in the truth.

Am I exaggerating? Not at all! Here is an example for you. Mathematics and science go hand in hand. A mathematician discovered that a widely accepted mathematical theory is contradictory in its assumptions and provided evidence to that effect. His discovery however was so problematic to his peers that he was killed; and his crime—*"__having produced an element in the universe which denied the...doctrine that all phenomena in the universe can be reduced to whole numbers and their ratios__"*. Here is the rest of the passage for you:

Ancient Greece

The first proof of the existence of irrational numbers is usually attributed to a __Pythagorean__ (possibly __Hippasus of Metapontum__) who *probably discovered them while identifying sides of the <u>pentagram</u>. The then-current Pythagorean method would have claimed that there must be some sufficiently small, indivisible unit that could fit evenly into one of these lengths as well as the other. However, Hippasus, in the 5th century BC, was able to deduce that there was in fact no common unit of measure, <u>and that the assertion of such an existence was in fact a contradiction</u>. He did this by demonstrating that if the <u>hypotenuse</u> of an <u>isosceles right triangle</u> was indeed <u>commensurable</u> with a leg, then that unit of measure must be both odd and even, which is impossible. His reasoning is as follows:*

- *Start with an isosceles right triangle with side lengths of integers a, b, and c. The ratio of the hypotenuse to a leg is represented by c:b.*
- *Assume a, b, and c is in the smallest possible terms (i.e. they have no common factors).*
- *By the <u>Pythagorean theorem</u>: $c^2 = a^2+b^2 = b^2+b^2 = 2b^2$. (Since the triangle is isosceles, a = b).*
- *Since $c^2 = 2b^2$, c^2 is divisible by 2, and therefore even.*
- *Since c^2 is even, c must be even.*
- *Since c and b have no common factors, and c is even, b must be odd (if b were even, b and c would have a common factor of 2).*
- *Since c is even, dividing c by 2 yields an integer. Let y be this integer (c = 2y).*
- *Squaring both sides of c = 2y yields $c^2 = (2y)^2$, or $c^2 = 4y^2$.*
- *Substituting $4y^2$ for c^2 in the first equation ($c^2 = 2b^2$) gives us $4y^2 = 2b^2$.*
- *Dividing by 2 yields $2y^2 = b^2$.*
- *Since y is an integer, and $2y^2 = b^2$, b^2 is divisible by 2, and therefore even.*
- *Since b^2 is even, b must be even.*
- *However, we have already asserted that b must be odd, and b cannot be both odd and even. This contradiction proves that c and b cannot both be integers, and thus the existence of a number that cannot be expressed as a ratio of two integers.[9]*

<u>Greek mathematicians</u> termed this ratio of incommensurable magnitudes alogos, or inexpressible. Hippasus, however, was not lauded for his efforts: according to one legend, he made his discovery while out at sea, and <u>was subsequently thrown overboard by his fellow Pythagoreans</u> "...for having produced an element in the universe which denied the...doctrine that all phenomena in the universe can be

reduced to whole numbers and their ratios.[10] *Another legend states that Hippasus was merely exiled for this revelation. Whatever the consequence to Hippasus himself, his discovery posed a very serious problem to Pythagorean mathematics, since it shattered the assumption that number and geometry were inseparable–a foundation of their theory.*

The discovery of incommensurable ratios was indicative of another problem facing the Greeks: the relation of the discrete to the continuous. Brought into light by Zeno of Elea, who questioned the conception that quantities are discrete and composed of a finite number of units of a given size. Past Greek conceptions dictated that they necessarily must be, for "whole numbers represent discrete objects, and a commensurable ratio represents a relation between two collections of discrete objects."[11] However Zeno found that in fact "[quantities] in general are not discrete collections of units; this is why ratios of incommensurable [quantities] appear….[Q]uantities are, in other words, continuous."[11] What this means is that, contrary to the popular conception of the time, there cannot be an indivisible, smallest unit of measure for any quantity. That in fact, these divisions of quantity must necessarily be infinite. For example, consider a line segment: this segment can be split in half, that half split in half, the half of the half in half, and so on. This process can continue infinitely, for there is always another half to be split. The more times the segment is halved, the closer the unit of measure comes to zero, but it never reaches exactly zero. This is just what Zeno sought to prove. He sought to prove this by formulating four paradoxes, which demonstrated the contradictions inherent in the mathematical thought of the time. While Zeno's paradoxes accurately demonstrated the deficiencies of current mathematical conceptions, they were not regarded as proof of the alternative. In the minds of the Greeks, disproving the validity of one view did not necessarily prove the validity of another, and therefore further investigation had to occur."

That an innocent man was either thrown overboard and drowned or exiled for his brilliance is troubling, but it helps to establish the fact that the world would prefer to live in darkness and in total ignorance than to be brought to the glaring truth that exposes everybody's inadequacy and possible condemnation.

This is the same exact thing that is going on in the world today. A greater percent of the human population would rather live blissfully holding out hope for science to triumph over the Bible than to allow themselves to be talked into repenting and seeking God's righteousness so they may receive God's grace and salvation.

Many have seen what came out of science over time so they readily believe the scientists that everything else would be answered if enough time is allowed. These people are ready to wait it out in the hope that science would continue to deliver as it had already done. But that is only playing into the devil's hands.

Satan understands the human mind more than mankind understands its own mind. That is why the human mind is the devils playground, in which the devil wins much more than he loses. He knows that mankind is a sucker to repeatability; if it worked before, it will work again and again. Although mankind has learned that it does not always continue to work the same way, but it is ready to wait it out until constant failures replace constant successes.

All the celebrated advances of science started with hypothesis that over long periods of time were proven and became theories. And many of those theories have led to amazing advances to man's quality of life. That being the case, we must take every hypothesis seriously since it could lead to another breakthrough. When the hypothesis for evolution came along with all the signs surrounding it appearing to point to the truth of the hypothesis, the scientists got excited and were willing to expand on it.

These expansions and embellishments became substantial enough and very long period of time had passed and no one was able to prove the hypothesis incorrect, the hypothesis then became a theory and quickly received wide acceptability in the world of science. Although the world at large still does not have enough confidence in evolution as a substitute for God, the world does not mind the subject being taught at schools because the world largely believes that there is more truth to the theory than fabrication.

And since science is all about improved products and improved ways of life and the management of society, government totally bought into science and promotes it unconditionally because of science's great benefits to government and society. And as long as science continued to benefit society and government, society became more apt to disbelieve the church than to disbelieve science, especially since large segments of the church dart all over the place in their claims of the same Biblical truth—which incidentally is another work of the devil. Not only does the devil manufacture lies and spreads it within human societies, he also infiltrates the truth within the church of God.

And since a great part of the process in science is continually incorporating improvements, mankind has grown accustomed to that process and is willing to allow whatever time scientist project to get it right. After all scientists are human beings and are not perfect. Scientists look more attractive to mankind than the Christian preachers because, while preachers stick to the same message that has been around since the world began, science and the scientists are constantly generating new thrills and excitement.

And this is simply the devils grand scheme! God's message is the same yesterday, today and forever more; and it is not for thrill seekers. It is a blue print for life—the right kind of life, the eternal life. The devil has simply got the world where he wanted the world to be—believe that the earth and the universe took billions of years to become what they are today and that they will be around for billions more years and that mankind has many years to find solutions to all of its problems.

And that is the devil's big deception! The biggest irony of this deception is that the world societies who have advanced in this current world and believe that they have life all figured out are the ones who would end up losing everything in their arrogance and ignorance of God and the Bible. The entire mankind belongs to God. Every single soul that walks the face of the earth is God's. And God is not a respecter of persons. He gives His grace to all without discrimination and He expects all to receive His grace and be saved. Anyone who decides to disregard God's truth only has himself or herself to blame. The unfortunate thing is that so many people like to go where everybody else is going. So in this regard, they will follow the skeptics into their doom for electing to simply accept the evidence presented to them without their own examination.

Like God said to Prophet Elijah, "I have seven thousand men in Israel who have not bowed their heads to Baal"; God continues to have in the current world so many people of different nationalities and convictions who have not bowed to false science and the deception of the devil. That you live in a country where most people strongly believe everything science says does not mean you should join them.

If you become like them, it is your fault, not your society's fault. Societies are but a fading entity. In the end, they will all varnish and each and every human being will answer for himself or herself. The society will not be there that day to lend support in your defense. It will simply be you and you alone!

None of us even have enough mental capacity to comprehend everything God has laid out for humanity in the Bible. Humanity as a whole has not developed enough comprehension for the science behind everything God created in the universe. To attest to the truth of this claim yourself, you may ask yourself this question: "How much do I really know about the world and the universe?"

If you do not know where to start, start with researching into what it may take for you to go into a new professional career—the years of study needed; or even a new specialty in your current field. Then think about all the other professional careers that come into your mind.

If you answer truthfully, you will discover that your total current knowledge of the world, the universe and all the professional specialties, is only a small fraction of one percent. And you believe that is enough to challenge God and His infinite wisdom?

Or that less than one percent makes you an expert in the things God says about His world? The best way to apply your knowledge of the world and the universe to your greatest benefit is to study the Bible and learn from God. He is still teaching mankind today and He will be only too happy to take you under His wings. Try Him and make your own determination. Don't just take my word for it.

God created you with everything you need to make your own choices. He continues to talk to you and interact with you every moment in your life. And it is not hard to sense Him reaching out to you. You just have to be hungry for an intimate relationship with Him and His Christ, and act when He tells you to.

Unfortunately for the people who are willing to give evolution enough time to produce concrete proof, God will not allow mankind that kind of long period before His judgment of all humanity. Therefore those who believe that evolution would one day prove the Bible incorrect would run out of time and end up in hell and eternal regrets.

Without the Bible and the unsearchable truth it contains—and of course, the God who continually controls everybody and everything in the universe—humanity would not have advanced, but rather would have regressed and self-destruct. The Bible says:

*"He has made everything beautiful in its time. **He has also set eternity in the human heart;** yet no one can fathom what God has done from beginning to end."* (*Ecclesiastes 3:11*).

Man cannot exist and progress without the spirit in man—which is "God in us". Out of love, God created mankind, and out of mercy, He redeems mankind! That's why the Bible says: *"because judgment without mercy will be shown to anyone who has not been merciful. **Mercy triumphs over judgment.**"* (James 2:13)

And it also says: *"**Above all, love each other deeply, because love covers over a multitude of sins.**"* (1 Peter 4:8).

And man is continually controlled by spirit—either the Spirit who is God or the evil spirit; there is no other mode of life coordination for humanity. Your "no" to God is your invitation and a welcome to the devil. There is no other way around this. Man can never do anything out of his own power because he has none. His every single action is controlled by either the good Spirit or the bad spirit.

Evolution is a fantasy. The facts which evolution claims as pointing to natural selection is nothing but the "single thread" of God which God puts in various classes of creation as a sign of His singleness; such as the biological cell in animals and the human body (not the human person who is spirit); the atomic number in the elements of the periodic table; and the wavelength in electromagnetic (light) spectrum.

The Earth Hangs on Nothing:

The Bible says that the earth hangs on nothing because the Spirit of God was holding it in position. God said that in Genesis 1:2 right before God dawned His light on the earth and TIME started. The earth was the dead center of all of God's creation. Everything God created, He created around the earth. In the beginning the earth was submersed in water. Then everything else God created started to happen around the earth, filling out in all directions.

The Bible says:

"Who has helped you utter these words?
And whose spirit spoke from your mouth?

[5] *"The dead are in deep anguish,*
those beneath the waters and all that live in them.
[6] *The realm of the dead is naked before God;*
Destruction lies uncovered.
[7] *He spreads out the northern skies over empty space;*
__*he suspends the earth over nothing.*
[8] *He wraps up the waters in his clouds,*
__*yet the clouds do not burst under their weight.*
[9] *He covers the face of the full moon,*
__*spreading his clouds over it.*
[10] *He marks out the horizon on the face of the waters*
__*for a boundary between light and darkness.*
[11] *The pillars of the heavens quake,*
__*aghast at his rebuke.*
[12] *By his power he churned up the sea;*
__*by his wisdom he cut Rahab to pieces.*
[13] *By his breath the skies became fair;*
__*his hand pierced the gliding serpent.*
[14] **And these are but the outer fringe of his works;**
how faint the whisper we hear of him!" (Job 26:4-14)

Chapter 8

CONSTANTS: SCIENCES BIGGEST BLUNDER & THE DEVIL'S GREATEST DECEPTIVE TOOL!

Scientific concepts like "gravity" are science's biggest blunder and the devil's best deceptive tools! Such concepts often take the appearance of irrefutability; hidden inside a constant that is continuously adjusted for improved accuracy. Gravitational constant is one such culprit and here is a little internet excerpt about it:

"Though Newton's Principia theorized the presence of the gravitational constant, it did not answer the question of the mathematical value of G. More than 70 years after Newton's death, a brilliant and fascinatingly eccentric <u>scientist</u> named Sir Henry Cavendish inherited a machine meant to measure the density of the Earth. The machine was the design of another scientist, Reverend John Michell, who died before he could complete his experiments. The fabulously complex machine, which was supposedly so sensitive it needed to be observed in operation from another room to avoid contaminating the results, helped produce not only the density results desired, but also led to future calculations of the gravitational constant.

Cavendish's calculations were not exactly correct, but even with 21st century technology, <u>the gravitational constant remains one of the most difficult physical constants to measure. Scientists have revised the calculations several times throughout the interim centuries, arriving in 2006 at a widely-accepted mathematical expression of G= 6.667428 X 10^{-11} m^3 kg^{-1} s^{-2}, where M=length in meters, kg=mass in kilograms, and s=time in seconds. With centuries of recalculation behind them and the potential for future centuries filled with more refinements, most scientific explanations add that this equation should still include some margin for error."</u>

Newton's first law of universal gravitation states that the

force of **gravity** between two masses is directly proportional to the product of the two masses and inversely proportional to the square of the distance between them, or mathematically: $F=G(m_1m_2/d_2)$, where G is a constant.

Using the equation, $g=GM/R^2$, the acceleration of gravity of different objects in space can be calculated. In the equation, g is gravity, G is the **gravitational constant,** **R** is the radius of the planet, and M is the mass of the planet, meaning that planets with more mass have a greater acceleration of gravity than planets with less mass.

A constant in a scientific theory does not have to tell anyone decisively what is contributing the constant. As long as the value appears to be correct by producing repeatable results, the constant is in until someone comes up with a scenario that significantly throws off its value. And a constant does not have to be a real number either; it can be imaginary.

Why is the use of constant widely accepted in science? Constants help in the abstract thinking process. They help to isolate certain things in a logical field or downplay their effect so that the effect of others would rise in prominence and lead to better understanding of the logic. Scientists often determine that some constants are good approximations of the real thing and they retain the constants.

And when these approximated situations are widely accepted by the scientists, they are disseminated as scientific facts which more often are accepted by society in the place of the Biblical truths. And replacing the Biblical truths automatically diminishes the importance of God in our lives. Even Christians come to accept these approximated situations as facts and begin to permit themselves to question the absoluteness of the claims of the Bible and God Himself. Science induces backsliding; and only a person of strong faith can resist its destructive force against his faith.

Let us look at a classical effect of these scientific constant on faith in God. Here is the scientific definition of gravity:

What is gravity?

1. *Gravity is a force pulling together all matter (which is anything you can physically touch). The more matter, the more gravity, so things that have a lot of matter such as planets and moons and stars pull more strongly.*

    Mass is how we measure the amount of matter in something. The more massive something is, the more of a gravitational pull it exerts. As we walk on the surface of the Earth, it pulls on us, and we pull back. But since the Earth is so much more massive than we are, the pull from us is not strong enough to move the Earth, while the pull from the Earth can make us fall flat on our faces.

    In addition to depending on the amount of mass, gravity also depends on how far you are from something. This is why we are stuck to the surface of the Earth instead of being pulled off into the Sun, which has many more times the gravity of the Earth.

2. *Essentially, gravity is an attractive force between objects. Most people are familiar with gravity as the reason behind things staying on the Earth's surface, or "what goes up, must come down," but gravity actually has a much vaster significance. <u>Gravity is responsible for the formation of our Earth and all other planets and for the movement of all heavenly bodies. It is gravity that makes our planet revolve around the Sun, and the Moon revolve around the Earth.</u>*

    Though humans have always been aware of gravity, there have been many attempts to accurately explain it throughout the years, and theories must regularly be improved upon to account for previously unconsidered aspects of gravity.

Aristotle was one of the first thinkers to postulate the reason for gravity, and his and other early theories relied on a geocentric model of the universe, with the Earth at its center. Galileo, the Italian physicist who made the first telescopic observations supporting a heliocentric model of the solar system, with the Sun at the center, also made strides in the theory of gravity around the turn of the 17th century. He discovered that objects of varying weights fall towards the Earth at the same speed.

In 1687, English scientist Sir Isaac Newton published his law of universal gravitation, which is still used to describe the forces of gravity in most everyday contexts. Newton's first law states that the force of gravity between two masses is directly proportional to the product of the two masses and inversely proportional to the square of the distance between them, or mathematically: $F=G(m1m2/d2)$, where G is a constant.

Using the equation, $g=GM/R^2$, the acceleration of gravity of different objects in space can be calculated. In the equation, g is gravity, G is the gravitational constant, R is the radius of the planet, and M is the mass of the planet. Doing the calculations, physicists have determined that the acceleration of gravity on Jupiter is approximately $85.3ft/s^2$ ($26m/s^2$). Pluto, on the other hand, has a value of $2 ft/s^2$ ($0.61m/s^2$). You can see that planets with more mass have a greater acceleration of gravity than planets with less mass.

If the world was a vacuum, these values would represent real life. On the moon, the air is a vacuum, and so objects fall to the ground at the acceleration of the moon's gravity. On Earth, however, we have air resistance--the force of air pushing against an object as it falls. This is the reason that a feather floats down to Earth while a bowling ball plummets, even though gravity is acting upon both objects equally. In order to accurately calculate the speed at which an object falls, air resistance must be accounted for

3. *Gravity is a very important force. Every object in space exerts a gravitational pull on every other, and so gravity influences the paths taken by everything traveling through space. It is the glue that holds together entire galaxies. It keeps planets in orbit. It makes it possible to use human-made satellites and to go to and return from the Moon. It makes planets habitable by trapping gasses and liquids in an atmosphere. It can also cause life-destroying asteroids to crash into planets.*

Let me summarize the whole claim for you.

- Gravity is an attractive force which every mass exerts over every other mass: we pull at the earth and the earth pulls back at us, and everything pulls at everything else.
- Gravity holds every mass together so that it remains a discrete entity.
- Gravity is what keeps bodies in orbit and causes them to rotate and revolve around a larger body.
- In essence, in their understanding, gravity is the stabilizing force that keeps all things in the universe in harmony, thereby eliminating the need for God or his presence.

Now let us look at the truth of the Bible with regard to gravity:

Genesis 1:2—*"Now the earth was formless and empty, darkness was over the surface of the deep, and the Spirit of God was hovering over the waters."*

This is before time existed. The dawn of light on the earth is the beginning of time in the universe at large, since the universe was not created until Day 4 of creation. Day 1 of creation marked the first day ever in the entire universe.

So before time existed, God created the Earth as a hot, molten mass (Job 38:14) submerged inside a deluge of Water (Genesis 1:2). Dense water vapor (Job 38:9) rose from the quenching of the hot mass (the molten earth) and caused thick darkness over the surface of the water which was covering the earth (Genesis 1:2). This darkness symbolized the darkness of the heart that would soon after creation descend on mankind and the world which can only be removed through God's grace. It must be inferred then that **matter** was created before **time, space** and the **universe**.

The statement *"**and the Spirit of God was hovering over the waters**"* directly translates into "and the Spirit of God was hovering over the waters holding everything together; and in position." And to be sure that this is what the Spirit of God was doing and is still doing throughout the earth and the universe today, see the following passage from the Bible:

"The Son is the image of the invisible God, the firstborn over all creation. ¹⁶ For in him all things were created: things in heaven and on earth, visible and invisible, whether thrones or powers or rulers or authorities; all things have been created through him and for him. ¹⁷ He is before all things, and in him all things hold together." (Colossians 1:15-17).

The statement from the passage *"**in him all things hold together**"* means that in Jesus Christ, all things are held together. Therefore if all things are held together in Him, what we widely refer to as gravity is the Spirit that is in Jesus Christ; who is also the Spirit of God.

We can then conclude that it is the Spirit of God that holds all of God's creation together, because we just read that from the Holy Bible. And this was why the Spirit of God was hovering over the waters in Genesis 1:2—to hold everything together,; and in their respective positions!

Let us all consider another statement from the passage: *"For in him all things were created: things in heaven and on earth, visible and invisible, whether thrones or powers or rulers or authorities; all things have been created through him and for him."*

If all things, visible and invisible on the earth and in the entire universe, were created in Christ Jesus, it then means that there is nothing in existence anywhere that exists outside of Jesus Christ. And that being the case, the Spirit of God who exists in Him exists inside and all over everything that exists everywhere—with no exception!

One may them ask, is God really that big? And that answer is unequivocally Yes, He is that Big! Look at the following passage from the Bible and consider what it says about the size of God:

"Who has measured the waters in the hollow of his hand,
or with the breadth of his hand marked off the heavens?
Who has held the dust of the earth in a basket,
or **weighed the mountains on the scales**
and the hills in a balance?
[13] Who can fathom the Spirit of the LORD,
or instruct the LORD as his counselor?
[14] Whom did the LORD consult to enlighten him,
and who taught him the right way?
Who was it that taught him knowledge,
or showed him the path of understanding?

[15] **Surely the nations are like a drop in a bucket;**
__*they are regarded as dust on the scales;*
__**he weighs the islands as though they were fine dust.**
[16] **Lebanon is not sufficient for altar fires,**
nor its animals enough for burnt offerings.
[17] Before him all the nations are as nothing;
they are regarded by him as worthless
and less than nothing.

[18] With whom, then, will you compare God?
To what image will you liken him?
[19] As for an idol, a metalworker casts it,
and a goldsmith overlays it with gold
and fashions silver chains for it.
[20] A person too poor to present such an offering
selects wood that will not rot;
they look for a skilled worker
to set up an idol that will not topple.

²¹ Do you not know?
  Have you not heard?
Has it not been told you from the beginning?
  Have you not understood since the earth was founded?
²² He sits enthroned above the circle of the earth,
  and its people are like grasshoppers.
He stretches out the heavens like a canopy,
  __and spreads them out like a tent to live in.
²³ He brings princes to naught
  __and reduces the rulers of this world to nothing.
²⁴ No sooner are they planted,
  __no sooner are they sown,
  __no sooner do they take root in the ground,
than he blows on them and they wither,
  __and a whirlwind sweeps them away like chaff.

²⁵ "To whom will you compare me?
  Or who is my equal?" says the Holy One.
²⁶ Lift up your eyes and look to the heavens:
  Who created all these?
He who brings out the starry host one by one
  and calls forth each of them by name.
Because of his great power and mighty strength,
  not one of them is missing.

²⁷ Why do you complain, Jacob?
  Why do you say, Israel,
"My way is hidden from the L<small>ORD</small>;
  my cause is disregarded by my God"?
²⁸ Do you not know?
  Have you not heard?
The L<small>ORD</small> is the everlasting God,
  the Creator of the ends of the earth.
He will not grow tired or weary,
  and his understanding no one can fathom.
²⁹ He gives strength to the weary
  and increases the power of the weak.
³⁰ Even youths grow tired and weary,
  and young men stumble and fall;
³¹ but those who hope in the L<small>ORD</small>
  will renew their strength.
They will soar on wings like eagles;

they will run and not grow weary,
they will walk and not be faint." (Isaiah 40:12-31)

And because God is that big, His Son Jesus Christ is that big also, another part of the same passage above says: *"For **God was pleased to have all his fullness dwell in him**, [20] and through him to reconcile to himself all things, whether things on earth or things in heaven, by making peace through his blood, shed on the cross." (Colossians 1:19-20).*

This is why Jesus Christ proclaimed in the gospel, *"I and the Father are one." (John 10:30)* And further, He expounded, *"Don't you believe that **I am in the Father, and that the Father is in me**? The words I say to you I do not speak on my own authority. **Rather, it is the Father, living in me, who is doing his work.**" (John 14:10)*

And if you are still not sure that everything created in the universe and on the earth exists in God and His Christ, here is yet another passage from the Bible with more explanation for you: *"There is one body and one Spirit, just as you were called to one hope when you were called; [5] one Lord, one faith, one baptism; [6] **one God and Father of all, who is over all and through all and in all.***

[7] But to each one of us grace has been given as Christ apportioned it. [8] This is why it says:

"When he ascended on high, he took many captives and gave gifts to his people."

[9] (What does "he ascended" mean except that he also descended to the lower, earthly regions? [10] He who descended is the very one who ascended higher than all the heavens, in order to fill the whole universe.)" (Ephesians 4:4-10).

And if you are curious the form of our existence within God and His Christ, look at this other passage from the Bible: *"'For **in him we live and move and have our being.**' As some of your own poets have said, 'We are his offspring.'" (Acts 17:28).*God does not only have control over you, He influences your thoughts and directs your actions to match your desires.

So in essence the Spirit of God at Genesis 1:2 was doing the following and still does them today:

- The Spirit of God held the earth in place—that is completely submerged inside the water, with no part of the earth anywhere close to the surface of the water.
- The Spirit of God completely maintained the water in a spherical shape so that the water covered completely the entire surface of the spherical earth submerged in it.
- The Spirit of God held the water together and prevented the water from spilling away, since the water and the submerged earth were both suspended on nothing—that is, holding them in orbit. Light has not even appeared at this point, nor had the sun been made to create the orbital energies science claims is supplied by the sun.
- The Spirit of God maintained the dense water vapor that rose from all around the surface of the water (spherical) in a spherical envelope around the water— ***"Who shut up the sea behind doors when it burst forth from the womb, when I made the clouds its garment and wrapped it in thick darkness"***—Job 38:8-9.
- The Spirit of God create all the materials as they exist today, imparting the various chemical and physical (and later biological) properties to each in their forms and states of existence.
- The Spirit of God is the chemical and covalent bonds that held the constituents of both the earth and the water together, in their independent existence.

Here are more passages from the Bible for your pleasure and enlightenment. Go through them and determine for yourself whether God knew what He was talking about or not. Is God not talking all serious science in all these?

Should we believe God or dismiss these uncontested truths? Apply all your scientific knowledge and expertise to the following and the rest of the scientific wonders God laid out for humanity in the Bible and see if you have enough understanding to challenge God:

Job 38:4-7 *"Where were you when I laid the earth's foundation? Tell me, if you understand. Who marked off its dimensions? Surely you know! Who stretched a measuring line across it? On what were its footings set, or who laid its cornerstone—while the morning stars sang together and all the angels shouted for joy?"*

Job 38:8-11 *"Who shut up the sea behind doors when it burst forth from the womb, when I made the clouds its garment and wrapped it in thick darkness, when I fixed limits for it and set its doors and bars in place, when I said, 'This far you may come and no farther; here is where your proud waves halt'?"*

Job 38:14 *The earth takes shape like clay under a seal; its features stand out like those of a garment."*

Job 38:32-33 *"Can you bring forth the constellations in their seasons... Do you know the laws of the heavens? Can you set up God's dominion over the earth?"*

Isaiah 45:11-12 *"This is what the LORD says—the Holy One of Israel, and its Maker: Concerning things to come, do you question me about my children, or give me orders about the work of my hands? It is I who made the earth and created mankind on it. My own hands stretched out the heavens; I marshaled their starry hosts."*

Proverbs 3:19-20 *"By wisdom the Lord laid the earth's foundations, by understanding he set the heavens in place; by his knowledge the watery depths were divided, and the clouds let drop the dew.*

<u>Amos 9:6</u> ***"He builds his lofty palace in the heavens and sets <u>its</u> foundation on the earth; he calls for the waters of the sea and pours them out over the face of the land— the Lord is His name."*** *(Amos 9:6)*

Let us summarize the two claims for gravity: human science's definition for gravity and God's definition for gravity. The world, through worldly wisdom, has determined and continues to maintain that:

- Gravity is an attractive force which every mass exerts over every other mass: we pull at the earth and the earth pulls back at us, and everything pulls at everything else.
- Gravity holds every mass together so that it remains a discrete entity.
- Gravity is what keeps bodies in orbit and causes them to rotate and revolve around a larger body.
- In essence gravity is the stabilizing force that keeps all things in the universe in harmony, thereby eliminating the need for God or his presence.
- Gravity involves a constant, gravitational constant, whose value— **$G = 6.667428 \times 10^{-11} \ m^3 \ kg^{-1} \ s^2$, where M=length in meters, kg=mass in kilograms, and s=time in seconds**—has been corrected so many times over the centuries, and still requires that it continue to undergo these adjustments, to make it produce accurate values in all of its applications.

And God, through the Bible, on the other hand, maintains that gravity is simply the Spirit of God at work in everything that God created on the earth and in the entire universe, visible and invisible, known and unknown. The Bible tells us that the Spirit of God was hovering over the waters in Genesis 1:2 and the Bible did not tell us that this Spirit of God ever lifted from the waters and the earth any time during the six days of creation or afterwards, because God remains in everything He created and all remained in Him; and that God continues to hold everything together and control and manage everything— **"one God and Father of all, who is over all and through all and in all,"** *(Ephesians 4:6),* leading to the following conclusion:

God's Equation: God is infinite energy, plus the sum total of all the energies that exist everywhere in the universe, and on the earth, visible and invisible; thermal, nuclear, sonic, light, magnetic, electrical, chemical, potential, hydrostatic, and all the other forms of energy known and unknown. And all these are summarized in the following elementary equation as:

$$E_{God} = E^{\infty} + \sum_{n=1}^{\infty} E_n$$

Where:

E_{God} = God's Energy;

E^{∞} = Infinite Energy;

E_n = Total Energy of its kind in the universe;

n = number of energy types available in the universe, known and unknown; visible and invisible.

So, who should we believe, God who created everything out of nothing and still manages and controls them or human beings like us who believe that God does not exist and devises fancy ways to obscure the wisdom of God by covering up their assumptions and approximations and fancy maneuvers that trumps most of mankind intellectually and exclude them from the debates?

The reason gravitational constant has taken this long to fine tune is because the assumptions made as to what is contributing the constant is not entirely true. Take pi for instance. Although that constant is the ratio of the circumference of a circle to the diameter of that circle, pi can be arbitrarily replaced in an equation with a ratio of integers such as 22/7 to produce an approximate result. Yet 22/7 explains absolutely nothing about a circle. They are simply random numbers picked to approximate a real physical entity whose value is known.

The Bible infers that as these bodies raced away in every direction from around the earth, God arranged them in clusters to form galaxies; set the galaxies and the individual bodies within each galaxy in orbits and their inter-relationships; and set their motions and functions.

While God was creating the universe and dealing with these extreme energies, He protected the fragile earth and its infrastructure from being vaporized by the excessive energies generated in the Big Bang; thereby exhibiting the limitlessness of His capacity—He is God and there is no other!

Galileo is widely hailed for postulating that the earth and the other planets are revolving around the sun, thereby making the sun the center of our solar system, and the world quickly dismisses those who believed that the earth is the center of the universe. The sun only became the center of our solar system after the sun was created by God on Day 4 of creation.

The truth remains that the earth is the center of the universe. In the beginning, the earth—and the water surrounding it—was the only thing is orbit because there was no universe at that point. The earth remained the only body in orbit for three full days before the sun, the moon, the stars and the universe were created through the Big Bang; and the earth began to rotate around the sun.

And it is tradition with God throughout the Bible to carry over everything He started without losing one beat. Secondly, our solar system being extremely small in size when compared to the whole universe still makes the earth the center of the universe. With God, it's all about common threads. God's common thread on the animal world got Charles Darwin and all the evolutionists all twisted and out of sync, causing them to extrapolate minor variations into non-realities and confusing the entire human society with them, thereby denying people the grace that was given freely to all, and thereby hurtling them towards the eternal lake of fire.

Here is God's confidence about His creation activities. As God was crafting His masterpiece, the angels watched and celebrated: Here is the Scripture: *"Where were you when I laid the earth's foundation? Tell me, if you understand. Who marked off its dimensions? Surely you know! Who stretched a measuring line across it? On what were its footings set, or who laid its cornerstone—while the morning stars sang together and all the angels shouted for joy? (Job 38:4-7).*

It has become increasingly common for scientists to seek a constant to make up for their failures in establishing predicted correlations.

One of the biggest hot-button topics the scientific world is currently pursuing is the search for the "god particle"—the Higgs Boson particle which scientist is responsible for imparting weight to all other particles that have mass in the universe.

A particle believed to be the Higgs Boson particle which was recently detected in their search was analyzed, and the result suggests just the opposite of the real and certifiable observation—that the universe is constantly expanding. Their result suggested that the universe, shortly after the Big Bang, would have imploded and disappeared. Here is that article for you:

"Say What? Higgs Boson Theorist Claims Universe Shouldn't Exist

The universe shouldn't exist — at least according to a new theory.

Modeling of conditions soon after the Big Bang suggests the universe should have collapsed just microseconds after its explosive birth, the new study suggests.

"During the early universe, we expected cosmic inflation — this is a rapid expansion of the universe right after the Big Bang," said study co-author Robert Hogan, a doctoral candidate in physics at King's College in London. "This expansion causes lots of stuff to shake around, and if we shake it too much, we could go into this new energy space, which could cause the universe to collapse."

*Physicists draw that conclusion from a model that accounts for the properties of the newly discovered **Higgs boson** particle, which is thought to explain how other particles get their mass. Faint traces of gravitational waves formed at the universe's origin also inform the conclusion.*

Of course, there must be something missing from these calculations.

"We are here talking about it," Hogan told LiveScience. "That means we have to extend our theories to explain why this didn't happen."

How the Big Bang went bang

*One possible explanation holds that during the fiery flash after the primordial Big Bang explosion, matter raced outward at breakneck speeds in a process known as **cosmic inflation**. This bent and squeezed space-time, creating ripples known as gravitational waves that also twisted the radiation that passed through the universe, Hogan said.*

Though those events would have occurred 13.8 billion years ago, a telescope at the South Pole known as the Background Imaging of Cosmic Extragalactic Polarization, or BICEP, recently detected the faint traces of cosmic inflation in the background microwave radiation that pervades the universe: in particular, characteristic twisted or curled waves called the B-mode pattern. (Other scientists have questioned the findings, saying the results may just be from dust in the Milky Way.)

How did the universe begin?

Gravity wasn't the only force at play in the early universe. <u>A ubiquitous energy field, called the Higgs field, permeates the universe and gives mass to the particles that trudge through the field.</u> Scientists found the telltale sign of that field in 2012, when they discovered the Higgs boson and then determined its mass. <u>[6 Implications of Finding a Higgs Boson Particle]</u>

With a greater understanding of cosmic inflation's properties and the Higgs boson mass, Hogan and his colleague, Malcolm Fairbairn, who is also a physicist at King's College London, tried to recreate the conditions of cosmic inflation after the Big Bang.

What they found was bad news for, well, everything. The newborn universe should have experienced an intense jittering in the energy field, known as <u>quantum fluctuation</u>. Those jitters, in turn, could have disrupted the Higgs field, in essence rolling the entire system into a much lower energy state that would make the collapse of the universe inevitable.

<u>Missing ingredient</u>

<u>So if the universe shouldn't exist, why is it here?</u>

<u>"The generic expectation is that there must be some new physics that we haven't put in our theories yet, because we haven't been able to discover them," Hogan said.</u>

<u>One leading possibility, known as the theory of supersymmetry, proposes that there are superpartner particles for all the currently known particles, and perhaps more-powerful particle accelerators could find these particles, Hogan said.</u>

<u>But the theory of cosmic inflation is still speculative</u>, and some physicists hint that what looked like primordial gravitational waves to the BICEP telescope may actually be signals from cosmic dust in the galaxy, said Sean Carroll, a physicist at the California Institute of Technology and author of "The Particle at the End of the Universe:

How the Hunt for the Higgs Boson Leads Us to the Edge of a New World."

If the details of cosmic inflation change, then Hogan and Fairbairn's model would need to adapt as well, Carroll told LiveScience. Carroll was not involved in the study.

This isn't the first time that physicists have said the <u>Higgs boson spells doom for the universe</u>. Others have calculated that the Higgs boson's mass would lead to a fundamentally unstable universe that could end apocalyptically in billions of years.

The mass of the Higgs boson, about 126 times that of the proton, turns out to be "right on the edge," in terms of the universe's stability, Carroll said. A little bit lighter, and the Higgs field would be much more easily perturbed; a little heavier, and the current Higgs field would be incredibly stable.

Hogan is presenting his findings Tuesday at the Royal Astronomical Society meeting in Portsmouth, England. The study was published May 20 in <u>Physical Review Letters</u>.

— Tia Ghose, LiveScience

To detect signs of an event that happened 13.8 billion years ago translates to our telescope at the South Pole picking up info from as far away as $8.09466048 \times 10^{22}$ miles away from the earth.

Our telescope at the south poles picked up an event from 8.09466048**e**22 miles away? Really? What did we do to isolate everything else that exists between us and this much distance away to ensure that none of them contributed to what we saw? What these scientists think is happening is really not what is happening.

Because of their ignorance of God, and their ignorance of the existence of the supernatural dimension that coexists with our world; they fail to realize how the powerful God of the universe can create situations from a few inches from their noses that look like they are coming from trillions of miles away from them, so that they may continue in their ignorant chase after nothing. This is a wake-up call, not an insult. God is God of love but He does not wait when it comes to illuminating ignorance.

"Supersymmetry, proposes that there are superpartner particles for all the currently known particles." Hooray! It looks like the human science is finally catching up with the Bible. There are indeed "superpartner particles" for every particle in the natural world and the natural universe—every particle already known to man, and those yet to be discovered by man!

Every natural particle has a corresponding spiritual "partner". Everything in nature: every human being; every plant and animal; every natural phenomenon; every human society; every human establishment; every classification in nature; and every quantifiable designation in the world and the physical universe; has a corresponding spiritual "partner". Everything in nature that has a discrete designation, and attribute, has a spiritual component that controls it.

Here is the Bible telling you so: *"There is one body and one Spirit, just as you were called to one hope when you were called;* [5] *one Lord, one faith, one baptism;* [6] **one God and Father of all, who is over all and through all and in all.**

[7] *But to each one of us grace has been given as Christ apportioned it.* [8] *This is why it says:*

"When he ascended on high,
he took many captives
and gave gifts to his people."

[9] *(What does "he ascended" mean except that he also descended to the lower, earthly regions?* [10] *He who descended is the very one who ascended higher than all the heavens, in order to fill the whole universe.)"* *(Ephesians 4:4-10)*

There is only one God, one Spirit, One Lord, who is the Father of all mankind, and the Lord of the earth and the entire universe. And this one God—who created everything in the first place—is *"over all and through all and in all."* This places God squarely in control of everything He created; through His Spirit.

"Supersymmetry" could, therefore, be explained as parallel worlds: one is the natural—the physical world we live in. And the other is the supernatural—the spiritual world that directly controls the one we live in.

These Parallel worlds exist, not side by side as in physically separated parallels. But rather one is contained completely within the other; in proportionate distribution. Virtually everything or every component in the natural is completely saturated with the spirit that controls it.

The parallelism of the two worlds is not in terms of geometrical congruency, but rather in terms of the one-to-one pairing of spirit to reality, for God's efficacious control of everything in our physical realism.

The supernatural world is fully aware of the natural world; and has to be aware of it, because it has dominion over it. The natural world, on the other hand, is oblivious of the supernatural world, since it could neither see nor detect the supernatural world.

And since the natural world is unaware of the supernatural world, the natural world dismisses, as madness, the claim that it is influenced and controlled by this parallel supernatural world.

That is the way the devil wants it, and he has operated that way since the incident at the Garden of Eden where Eve believed the serpent and rebelled against God. The darkness on the surface of the water in Genesis 1:2 symbolizes this darkness that fell on the human mind which prevents mankind from seeing spiritual truths. Mankind is ignorant of the dark powers of the devil; and the overwhelming corrupting influence of the darkness that has enveloped the human mind.

By so doing, the natural world readily plays into the hands of the "powers and principalities" that are truly in control of our world, and has been steering the world into all the wrong directions. That is why the Bible prophesies that there must be a time of reckoning for the natural world—a time of judgment by the true God of creation who made everything and gave His commands to the humans for the benefit of their salvation.

That is what the Bible is talking about when it speaks about powers and principalities in the heavenly realm. And the supernatural realm far surpasses the natural realm in power and authority, leaving the natural completely at the mercy of the spiritual powers.

Satan, who is spirit, stole the dominion of the earth from Adam and Eve in the natural; and remains in control of all human societies from that point forward. How does he stay in control? He appoints his princes as heads of every single human government, small or large; powerful or weak. His princes then control all human leadership by exerting their authority and influence over the human leadership, causing them to operate as directed by these spirits.

Humanity, or more specifically, scientists, have spent so much time, energy and resources, chasing after the Higgs Boson particle to confirm their claims that the earth, the universe and man came into existence through chance and time; only to end up with a conclusion that the universe should never have existed in the first place, but should have collapsed and disappeared if their theories were correct. And that is a bit of a set-back for the human science, but it will not be for long. We have sure ways of manipulating things until they support our theories—we could possibly come up with some constant of imaginary nature.

Just listen to these comments from some scientists with regard to the disappointing Higgs Boson conclusion:

"That means we have to extend our theories to explain why this didn't happen."

"The generic expectation is that <u>there must be some new physics that we haven't put in our theories yet, because we haven't been able to discover them.</u>"

Yes! All we have to do anytime we run into problem with our genius speculations is add a little here and take away a little there; and if that still doesn't work, throw in a constant and bam, we got it! Is this really the way to learn the truth? God can give us the truth if we obey Him and start all our inquiries with the Bible. Most of what we need is already written for us. What we are currently lacking is faith and a little humility.

Chapter 9

THE KINGDOM OF GOD

All mankind is created to be kingdom-minded --- the kingdom of God, that is. And that is why the Bible says: *"Fear God and obey his commandments, for this is the whole duty of man. For God will bring every deed into judgment, including every hidden thing, whether it is good or evil." (Ecclesiastes 12:13-14).* This is the most profound advice in all the Bible; and we all should take heed to it. We should all forget about our selfish personal desires and lofty ambitions and consider the real purpose of God for our creation because that is the only way any of us is going to find God's full favor.

Is it possible to find God's full favor? The answer is unequivocally yes! And that is God's desire for all mankind. Every parent desires the best for his/her children; and that does not start with mankind --- we inherited it from God just the same way we inherited all our other good qualities from Him: *"So God created man in His own image; in the image of God he created him; male and female he created them." (Genesis 1:27).*

The Bible tells us that we can do all things in Christ who strengthens us. And all things mean all things with no exception. That is God's original intention for mankind: to do all things in Him. The Bible tells us that God continued to fellowship with Adam and Eve every evening at the Garden of Eden until they disobeyed God and were both driven from the Garden.

This is similar to what God did when He first created the heavens and the earth: God's direct light shined on the earth for the first three days. And on the fourth day, God created the universe and set the sun, the moon and the stars in the universe to take His place as the light on the earth; to prevent Him from making Himself visible to mankind every day. In a similar fashion, God came down to the Garden of Eden every evening and spent time with Adam and Eve but when darkness descended onto their hearts, God pronounced His judgment on them and ended His daily visits with them.

God put His Spirit in every human being He created to give mankind life and understanding. And through this spirit in man, God gives discernment to man. But due to the darkness that descended onto mankind and led Adam and Eve to disobedience of God, thereby compromising man's holiness; man's spirit died and man had to rely on his mind to get through life in the world, and to search for God because his heart continues to yawn for God.

Whichever heart finds God and obeys God regains the spiritual discernment. And whichever heart continues to disobey God continues to lack discernment and remains blind. That is why Abel knew what would please God and sacrificed the best of his harvest to God. And God was pleased with Him. But Cain on the other hand, followed his mind and reserved the best of his harvest for himself and sacrificed his leftover to God. And God was displeased with Cain.

But because God's Will and His desire for mankind still remains the same due to God's immeasurable goodness and mercy, He gave mankind His grace by sacrificing the life of His Son Jesus Christ to empower man and help him regain his holiness so he could receive God's salvation.

Man's selfishness and intoxication for power, however, continue to drive man and keep his focus on the pursuit of pleasures and physical comfort at the expense of his spiritual wellbeing. And it has gotten to a point where most of mankind has forgotten that man's spiritual wellbeing depends on man's relationship with God.

When life is going well, we conveniently forget that our spiritual existence is something we have to assign time and effort to. We wait until calamities show up in our lives before we seek God. Man has excelled in every field of life but true spirituality, because our knowledge acquisition method is based on the examination of things we can affect and control. But when it comes to our spirituality, we realize that we are dealing with a Superior Being who cannot be manipulated but must be obeyed completely.

God is not a thing that we can post up and manipulate at will and dissect and observe. Those who are humble bow their heads and begin to learn from God. And those who are arrogant and interested in only exalting themselves dismiss the reality of God as mere fallacy and man's insecurity and a ploy to bolster one's waning courage.

While the former grew spiritual and received from God what is desirable to discern spiritual knowledge, the latter convinced themselves that superior mental knowledge will lead to the truth of God—if God truly exists. Unfortunately for them, the mind procures knowledge from man's immediate surrounding and past experiences.

But a man's spirit, receives universal knowledge from God from far beyond man's environment and mental grasp. Therefore, we are unintentionally limiting our learning potential when we exclude God from our lives. God reaches into time and space and gives to those who believe in Him knowledge of things that are yet to come into the world.

So when the Bible says that obeying God is the whole duty of man, is it an overstatement? No, it is not an overstatement. God made mankind so mankind will obey His every command. It is not only supposed to be the number one priority in everybody's life. It is the only duty we have!

Obeying God is serving God. And serving God is worshipping God. Therefore, worshipping God is the whole duty of every single human being.

We have come to regard the millions of man-made doctrines across the world's religious sphere as the worship of God; and are focusing all our efforts squarely on them. That brings our current Christian Faith to the same level as the Pharisees and the teachers of the Law at the time of Jesus Christ in Israel. The worship of God in today's churches has become a set of man-made religious doctrines, just like they were at that time.

Jesus Christ was so critical of those Jewish leaders. He blamed them for not following what Moses taught them that will get them into the kingdom of heaven; and also for blocking the worshippers who diligently seek God from entering, through their adulterated teaching and lifestyle.

From the very beginning, man has demonstrated a great inability to understand what God intended man's relationship with God to be. The purpose of man's entire life is to worship God; and it is not a selfish thing on the part of God. It stems from God's limitless love for man and the need to protect man from himself, because man has a self-destructive nature.

If you are curious as to how God expects us to worship Him continually and still be able pursue and maintain a gainful life, we will get to it in a minute. First I want to remind the reader that in the beginning, God was supposed to be responsible for all our provision and He did with Adam: He put together a lush garden and put Adam in it to dwell in. And Adam had everything he needed. But once sin came into man's life due to man's greed and unbelief, God kicked Adam and Eve out of the Garden of Eden to fend for themselves.

The best they did for themselves was cover themselves with leaves that they sew together. But God out of His unending love and great mercy and compassion, clothed them instead with animal skin and atoned for their sin with the blood of the animals. And God continued to be with them and in them and watched out for them because they could do nothing without God.

Although man through his own action has changed his destiny, God continued to provide for him but this time with some effort from man to demonstrate his desire to be fed and clothed and sheltered. And the very next generation of mankind realized it is their duty to honor God with the fruits of their labor: Cain with his crop harvest and Abel with his livestock.

But even from that event man demonstrated his self-centeredness and reluctance to follow the Will of God for his life. Abel gave from his best harvest but Cain gave from his leftovers and became murderously enraged when his offering was rejected by God.

Abel learned from God through his spirit what will make a pleasing sacrifice to God and offered it. Cain on the other hand could care less what God thinks of his sacrifice for whatever reason best known to him. However when God showed his dissatisfaction about what Cain offered to Him, Cain became jealous of his brother for giving to God what pleased God. Now he could not feel any comfort because of his indifference and selfishness.

Abel was able to respond in a way that totally satisfied God because he was conducting his life in a manner that kept him close to God and kept God interested in everything he was doing in his life. Abel was a spiritual man and maintained spiritual connection to God as God intended for man to do. Through this spiritual connection, Abel was able to know the mind of God and satisfy God's desire.

Cain on the other hand, had other issues to contend with in his life. He followed his physical and mental abilities more than he needed to and neglected his spiritual needs. In essence, He did not acknowledge God in all things. He pursued life on his own and exclusively followed the direction of his mind; thereby missing the whispers of God that fetches corrections from God at every turn in life.

Abel may have died in the hands of his brother but he had a fulfilling life. He was in the world but not of the world. He served God contentedly and received the rewards that are given by God for obedience and dedication. The Bible did not tell us how old he was when he died, but it was clear he died a noble member of the kingdom of heaven for the book of Hebrews and the gospel told us that Abel was a prophet of God and his blood is still speaking today.

The value to God of worship within the confines of the brick and mortar that we call church is directly proportional to the amount of the time we each spend within those confines. The time you spend at your place of worship every week does not make up for the time you spend on your day-to-day life's activities in which you are not worshipping God. All your time must constitute a service to God—that is your whole duty on in life. *(Ecclesiastes 12:13-14)*.

So if we spend two hours in church services every week that is only two (2) hours out of a possible eighty four (84) hours we are awake and should be worshipping God. We have then only given God a very small fraction (2/84) of what He demanded from us.

For those who serve other missions in the church, they probably spend on average, another 8 hours each week, bringing their total time of service to God to 10 hours/week. That is 10/84 of the time available to us for worshipping God. If we add time spent at home in prayers, we can add another 10 hours a week bringing our total to 20/84 hours per week This is still a far cry from continually serving God as the passage in Proverbs and other Scriptures commanded.

I used worship in this analysis because when people worship God, they do not have time or the opportunity to become disobedient to God. Therefore all the time we spend in worship could be counted as times we are obedient to God's commandments. That leaves us with 60 hours out of a possible 84 hours each week that we could be in disobedience of God, and we largely are, because we no longer has the priests and the church environments to continue to remind us about our obligation to be obedient to God always.

God knows that we have to work to provide for ourselves and our family. Does He expect us to be serving Him even when we should be working for a living? That is the point exactly! Everything we do in this life must be a service to God or we are in disobedience. Work and our other life activities that involve other people are exactly the events that test our obedience to God. If there is no one else even remotely involved in what any activity you are performing then there is little or no room for you to be in disobedience of God.

It is our interaction with other human beings that produce a lot of testing opportunities for our obedience of God. We mostly take our jobs for granted when it comes to obeying God and as such we err greatly in that area. And because we spend 40 out of 84 waking hours at work every week, the work place becomes a fertile ground to demonstrate our obedience to God. Any activities of ours that places us in interaction the most with other people are the most important activities for us to be "salt and light to the world". So our work places are up there in importance to our obedience to God.

For starters, we must view our jobs as gifts from God, because they are indeed. That is how we feed our families and cater for their every need. People are miserable when they could not find work. They live with great uneasiness and despair and lack. So our jobs and careers are great gifts from our loving Father who oversees everything in our lives.

At work, we should serve in whatever we are charged with as though we are providing a service to God; because we truly are. Others within the same business that we work for are depending on our professional contributions and cooperation to perform their own functions. And our collective contributions produce the profit that is used to pay our respective salaries and keep us employed. Therefore, it does not matter who the owners of these businesses are and their personal disposition, we must take our jobs very seriously as a service to God.

That is why the Bible commands everyone who is serving in any business outfit to serve the business wholeheartedly as if they were serving God Himself. Meaning that we must extend the same obedience and goodwill we show in our church services on Sundays to the people we work for and work with. It is all summed up in the following passages from the Bible. The passages are not about slavery, they are about just anyone who is providing their services to others for financial benefits. They are also for the people they work for:

"Slaves, obey your earthly masters with respect and fear, and with sincerity of heart, just as you would obey Christ. ⁶ Obey them not only to win their favor when their eye is on you, but as slaves of Christ, doing the will of God from your heart. ⁷ Serve wholeheartedly, as if you were serving the Lord, not people, ⁸ because you know that the Lord will reward each one for whatever good they do, whether they are slave or free.

⁹ And masters, treat your slaves in the same way. Do not threaten them, since you know that he who is both their Master and yours is in heaven, and there is no favoritism with him." (Ephesians 6:5-9)

"Slaves, obey your earthly masters <u>in everything</u>; and do it, <u>not only</u> <u>when their eye is on you and to curry their favor, but with sincerity of</u> <u>heart and reverence for the Lord</u>." (Colossians 3:22)

"Also, seek the peace and prosperity of the city to which I have carried you into exile. Pray to the LORD for it, because if it prospers, you too will prosper." (Jeremiah 29:7).

So in the 40 hours or so you spend at work each week, if you resolve to fully obey these commands of God at the work place—being obedient to your superiors, being courteous and helpful to your coworkers, being kind and considerate in your supervision and management of those under you, being courteous and considerate to the clients using your services, and squeezing in meditation and silent prayers in the course of the day—you can add another 40 hours to your weekly total of your service to God. This raises your total weekly worship time significantly to 60/84 possible worship hours. This is very impressive.

In the 40 hours you interact at work with your coworkers, your superiors and the clients/customers, you should remember these commands of God and make serious effort to apply them throughout the time you are performing your duties. Then all of that time will count as worship to God. That does not mean that if you slip up and momentarily go into disobedience that you have lost credit for that time.

God looks at your intent more than He looks at your success rate. It does not matter that you fail as long as you get right up, dust off, and march on. The Bible says that is the constant use of the commands God gave to us in the Bible that leads us to maturity in the faith. It may be harder in the beginning but before long, it becomes a second nature. Be willing and prepared to mesmerize others, and bring them closer to the Father of all mankind who has given you handsomely and waits to give to the rest of humanity who chooses to obey.

The important thing is to have the intention to be in obedience of God throughout the time you are at work, and diligently pursue it. And you need to actively remind yourself of that very important goal for you to be successful at it. Not only will you get the mark for being obedient to God, you will also receive God's reward for caring about the wellbeing of others—your coworkers and your clients.

It is through your success at work in staying in obedience to God as He commanded in these passages that the people you work for, and with, notice your Christ-like qualities and may seek to know how you are able to do it. This is why the Bible says:

"But in your hearts revere Christ as Lord. Always be prepared to give an answer to everyone who asks you to give the reason for the hope that you have. But do this with gentleness and respect, [16] keeping a clear conscience, so that those who speak maliciously against your good behavior in Christ may be ashamed of their slander. [17] For it is better, if it is God's will, to suffer for doing good than for doing evil. [18] For Christ also suffered once for sins, the righteous for the unrighteous, to bring you to God. He was put to death in the body but made alive in the Spirit. [19] After being made alive, he went and made proclamation to the imprisoned spirits— [20] to those who were disobedient long ago when God waited patiently in the days of Noah while the ark was being built. In it only a few people, eight in all, were saved through water, [21] and this water symbolizes baptism that now saves you also—not the removal of dirt from the body but the pledge of a clear conscience toward God. It saves you by the resurrection of Jesus Christ, [22] who has gone into heaven and is at God's right hand— with angels, authorities and powers in submission to him." (1 Peter 3:15-22)

Not only would people want to acquire some of your qualities, they would also give thanks to God for putting someone like you in their midst. Your character and wisdom is known and celebrated all over your work place as invaluable. This is an additional blessing for you. You have allowed Jesus Christ all access into your life and He is touching people everywhere you go; because if you successfully do this at work, it follows you everywhere else you go.

And this raises your total worship time to God to 84 out of 84 possible waking hours every week. And this is a sample of how one can be fully in compliance of *Ecclesiastes 12:13-14.*

Chapter 10

GODLINESS IS INFINITELY ...

Godliness is ...

Infinitely divine ("Is it not written in your Law, 'I have said you are "gods"'? ⁻John 10:34-38)

Infinitely supernatural

Infinitely human ("the whole duty of man" – Ecclesiastes 12:13)

Infinitely natural

Infinitely beyond religion

Infinitely praiseworthy

Infinitely humble

Infinitely humbling

Infinitely living

Infinitely good

Infinitely timeless

Infinitely demanding (except the **continuing debt** to love one another – Romans 13:8)

Infinitely peaceful

Infinitely favorable

Infinitely joyful

Godliness is ...

Infinitely blissful

Infinitely resourceful

Infinitely secure

Infinitely free

Infinitely priceless

Infinitely rewarding

Infinitely remarkable

Infinitely outstanding

Infinitely merciful

Infinitely graceful

Infinitely forgiving

Infinitely glorious

Infinitely loving

Infinitely giving

Infinitely caring

Infinitely inspiring

Infinitely exhilarating

Infinitely trusting

Infinitely reassuring

Infinitely satisfying

Godliness is ...

Infinitely supporting

Infinitely appreciating

Infinitely encouraging

Infinitely sharing

Infinitely wise

Infinitely adoring

Infinitely praiseworthy

Infinitely kind

Infinitely discerning

Infinitely soothing

Infinitely sensible

Infinitely hospitable

Infinitely friendly

Infinitely consoling

Infinitely majestic

Infinitely noble

Infinitely charitable

Infinitely memorable

Infinitely invaluable

Infinitely beneficial

Godliness is ...

Infinitely mellow

Infinitely luminous

Infinitely enlightening

Infinitely invigorating

Infinitely strengthening

Infinitely liberating

Infinitely unrestrictive

Infinitely abounding

Infinitely honorable

Infinitely applicable

Infinitely affable

Infinitely honest

Infinitely upright

Infinitely discrete

Infinitely sensible

Infinitely brotherly

Infinitely sisterly

Infinitely fatherly

Infinitely motherly

Infinitely clean

Godliness is ...

Infinitely wholesome

Infinitely uplifting

Infinitely cleansing (from sin or guilt)

Infinitely objective

Infinitely desirable

Infinitely protective

Infinitely consistent

Infinitely pure

Infinitely incorruptible

Infinitely uncompromising

Infinitely desirous

Infinitely mature

Infinitely effective

Infinitely energetic

Infinitely energizing

Infinitely noteworthy

Infinitely unpretentious

Infinitely unassuming

Infinitely uncondescending

Infinitely open

Godliness is ...

Infinitely fascinating

Infinitely mesmerizing

Infinitely captivating

Godliness is infinitely Christ. *{"I am the vine; you are the branches. If you remain in me and I in you, you will bear much fruit; apart from me you can do nothing." (John 15:5)}*

And finally, godliness is infinitely GOD. *{"On that day you will realize that I am in my Father, and you are in me, and I am in you." (John 14:20)}*

Chapter 11

MERCY TRIUMPHS OVER JUDGMENT!

"Mercy triumphs over judgment." What does God mean by this declaration? Is God condoning sin? Not at all! God is reaffirming what He told us in other places in the Bible. Judgment is reserved for God alone. No human being has the right to judge another human being. By the statement God is telling every one of us to continue to love and be merciful to everyone even when they have clearly fallen into sin. He did not say to embrace their sin or dismiss their sin as good. He warned us to be careful when helping anyone who was ensnared in sin or we could become overcome by the same sin.

So the right application of the statement is: Denounce sin as sin and stay as far away as possible from it. Pray to God to safeguard you from their sin. Then offer assistance to the person afflicted by the sin, praying to God that they may be redeemed from the sin. Lead the person to what is necessary to confess their sin and resolve never to do it again and mean it with all their heart, and continue to try and reach God throughout their ordeal.

Never for a second downplay what they have done or take it for granted that because God forgives all sins what they have done is really no big deal because you will bring judgment on yourself and will draw God's wrath. Sin must always be repudiated and kept away. Any downplaying of sin could be seriously consequential. We must always remember: God's way is higher than our way; and His mind is higher than ours.

Be careful how involved you become: Your responsibility is not to outline to anyone what part of their actions constitutes sin. That is what their conscience is for. If they do not recognize their disobedience of God as disobedience, you are playing with fire because you might be sucked in and burned. If their conscience is seared, there is nothing you can do about it. Only God can restore their conscience and make them feel things the right way again.

The Bible says: "The Gentiles who did not receive the law are a law for themselves because ..." So God still communicates with every human being through their spirit. It is not your duty or anyone else's to dissect sin and pronounce to the sinner what is wrong with their sin. The sinner already knows it because God is sure to get that point across to everyone who disobeys His commands. And that applies to everything God has commanded mankind in the Bible.

God said love one another so that the world will know that you belong to me. How would anybody measure love except by being obedient to all the commandments of God. If you choose to disobey any of God's commandments at any point, you are failing to love as God says to love. Obedience to God's commandments is the only way to measure anyone's love of God.

That is why Jesus Christ says: "If you love me, obey my commands." "I obey my Father's commands because I love the Father." "I obey everything the Father says."

Obeying all of God's commands bring us God's blessings and God's protection "If you obey my commands, my Father and I will come to you and build our house near you." So you want God's love, mercy, protection and blessings, obey all God's commandments. Do not say it is impossible to do so because you will not be doing it by your own powers; the Spirit of God will empower you to do it.

Anytime you wonder if you are doing it right, ask yourself this question: "Am I obeying everything God has commanded me to obey? Or am I picking and choosing what to obey and what not to obey because I do not like how God stated it?" You will immediately know your answer. You do not need to get a pastor to explain it to you.

Just like we receive human approval when we do things right, God communicates His approval and disapproval directly to our spirits for every one of our actions. We immediately know what we've done and how well it goes with God, except for those of us who have trained themselves to deny the obvious.

Unfortunately denying that the wrongs you do are wrongs do not change your wrongs to good. Wrong remains wrong whether we admit they are wrongs or not. God stripped all mankind of the right to judge one another because humans have the tendency to rationalize. And any wrong humanity could rationalize becomes right in our eyes: If anything applies to the rest of us, then it is not that bad because we all do it.

But we should all be reminded: God's way is higher than our way! Wrong is wrong and remains wrong till the end of the world. That is why God sent His Son Jesus Christ to shed His precious blood so He could atone for all our wrongs. Besides, the blood of Jesus Christ, nothing in the world or in heaven can atone for sins. And all sins must be atoned or they will not be forgiven by God.

"Mercy triumphs over judgment" is then infinitely human. It applies in every human situation. It is the only human approach to sin that is acceptable to God. And this is limited to one's ability, meaning everyone does not have the same capacity for mercy, but every single one of us must ascribe to it. Everyone's best effort in any given instance of mercy is a win for the kingdom of God, so we should all be encouraged to be merciful in every situation demanding it.

Look at the following passage from the Bible: *"And we urge you, brothers and sisters, warn those who are idle and disruptive, encourage the disheartened, help the weak, be patient with everyone.*

[15] Make sure that nobody pays back wrong for wrong, but always strive to do what is good for each other and for everyone else." (1 Thessalonians 5:14-15).

You can only offer the best of your abilities but you must continue to strive to do so. *"With this in mind, we constantly pray for you, that our God may make you worthy of his calling, and that by his power he may bring to fruition your every desire for goodness and your every deed prompted by faith." (2 Thessalonians 1:11)*

And by continuously striving to do what is good, your own abilities are amplified every time to make your efforts effective: *"For through the law I died to the law so that I might live for God. [20] I have been crucified with Christ and I no longer live, but Christ lives in me. The life I now live in the body, I live by faith in the Son of God, who loved me and gave himself for me. [21] I do not set aside the grace of God, for if righteousness could be gained through the law, Christ died for nothing!" (Galatians 2:19-21)*

I continue to live in the grace of God because it is God's righteousness—sacrificing His Son to atone for my sins and His Holy Spirit illuminating my path as I go through life—that guide me through everything I do in life. It is this righteousness of God that permits anyone to choose mercy over judgment on behalf of a brother or sister that fell into sin; and is truly repentant of his/her sin. Without repentance on the sinner's part, you are trying to exceed your authority and could fall harder than the one you are trying to rescue.

Helping to camouflage a sin is tantamount to promoting sin. Sin must be condemned as sin. Otherwise it will infect everyone near and far. To be merciful instead of passing judgment means to encourage, comfort and urge the fallen brother or sister to repent and dedicate themselves to a life that is worthy of God, who calls him/her into His kingdom and glory *(1 Thessalonians 2:12)*

Here some passages from the Bible to consider:

Live as Those Made Alive in Christ

"Since, then, you have been raised with Christ, set your hearts on things above, where Christ is, seated at the right hand of God. ² Set your minds on things above, not on earthly things. ³ For you died, and your life is now hidden with Christ in God. ⁴ When Christ, who is your life, appears, then you also will appear with him in glory.

⁵ Put to death, therefore, whatever belongs to your earthly nature: sexual immorality, impurity, lust, evil desires and greed, which is idolatry. ⁶ Because of these, the wrath of God is coming. ⁷ You used to walk in these ways, in the life you once lived. ⁸ But now you must also rid yourselves of all such things as these: anger, rage, malice, slander, and filthy language from your lips. ⁹ Do not lie to each other, since you have taken off your old self with its practices ¹⁰ and have put on the new self, which is being renewed in knowledge in the image of its Creator. ¹¹ Here there is no Gentile or Jew, circumcised or uncircumcised, barbarian, Scythian, slave or free, but Christ is all, and is in all.

¹² Therefore, as God's chosen people, holy and dearly loved, clothe yourselves with compassion, kindness, humility, gentleness and patience. ¹³ Bear with each other and forgive one another if any of you has a grievance against someone. Forgive as the Lord forgave you. ¹⁴ And over all these virtues put on love, which binds them all together in perfect unity.

¹⁵ Let the peace of Christ rule in your hearts, since as members of one body you were called to peace. And be thankful. ¹⁶ Let the message of Christ dwell among you richly as you teach and admonish one another with all wisdom through psalms, hymns, and songs from the Spirit, singing to God with gratitude in your hearts. ¹⁷ And whatever you do, whether in word or deed, do it all in the name of the Lord Jesus, giving thanks to God the Father through him."

Instructions for Christian Households

"Wives, submit yourselves to your husbands, as is fitting in the Lord.

[19] Husbands, love your wives and do not be harsh with them.

[20] Children, obey your parents in everything, for this pleases the Lord.

[21] Fathers,[c] do not embitter your children, or they will become discouraged.

[22] Slaves, obey your earthly masters in everything; and do it, not only when their eye is on you and to curry their favor, but with sincerity of heart and reverence for the Lord. [23] Whatever you do, work at it with all your heart, as working for the Lord, not for human masters, [24] since you know that you will receive an inheritance from the Lord as a reward. It is the Lord Christ you are serving. [25] Anyone who does wrong will be repaid for their wrongs, and there is no favoritism." (Colossians 3:1-25)

"I want you to know how hard I am contending for you and for those at Laodicea, and for all who have not met me personally. [2] My goal is that they may be encouraged in heart and united in love, so that they may have the full riches of complete understanding, in order that they may know the mystery of God, namely, Christ, [3] in whom are hidden all the treasures of wisdom and knowledge. [4] I tell you this so that no one may deceive you by fine-sounding arguments. [5] For though I am absent from you in body, I am present with you in spirit and delight to see how disciplined you are and how firm your faith in Christ is." (Colossians 2:1-5)

Spiritual Fullness in Christ

"So then, just as you received Christ Jesus as Lord, continue to live your lives in him, [7] rooted and built up in him, strengthened in the faith as you were taught, and overflowing with thankfulness.

[8] See to it that no one takes you captive through hollow and deceptive philosophy, which depends on human tradition and the elemental spiritual forces[a] of this world rather than on Christ.

⁹ For in Christ all the fullness of the Deity lives in bodily form, ¹⁰ and in Christ you have been brought to fullness. He is the head over every power and authority. ¹¹ In him you were also circumcised with a circumcision not performed by human hands. Your whole self ruled by the flesh[b] was put off when you were circumcised by[c] Christ, ¹² having been buried with him in baptism, in which you were also raised with him through your faith in the working of God, who raised him from the dead.

¹³ When you were dead in your sins and in the uncircumcision of your flesh, God made you[d] alive with Christ. He forgave us all our sins, ¹⁴ having canceled the charge of our legal indebtedness, which stood against us and condemned us; he has taken it away, nailing it to the cross. ¹⁵ And having disarmed the powers and authorities, he made a public spectacle of them, triumphing over them by the cross."
(Colossians 2:6-15)

Freedom from Human Rules

"Therefore do not let anyone judge you by what you eat or drink, or with regard to a religious festival, a New Moon celebration or a Sabbath day. ¹⁷ These are a shadow of the things that were to come; the reality, however, is found in Christ. ¹⁸ Do not let anyone who delights in false humility and the worship of angels disqualify you. Such a person also goes into great detail about what they have seen; they are puffed up with idle notions by their unspiritual mind. ¹⁹ They have lost connection with the head, from whom the whole body, supported and held together by its ligaments and sinews, grows as God causes it to grow.

²⁰ Since you died with Christ to the elemental spiritual forces of this world, why, as though you still belonged to the world, do you submit to its rules: ²¹ "Do not handle! Do not taste! Do not touch!"? ²² These rules, which have to do with things that are all destined to perish with use, are based on merely human commands and teachings. ²³ Such regulations indeed have an appearance of wisdom, with their self-imposed worship, their false humility and their harsh treatment of the body, but they lack any value in restraining sensual indulgence."
(Colossians 16-23)

"Finally, brothers and sisters, whatever is true, whatever is noble,

whatever is right, whatever is pure, whatever is lovely, whatever is admirable—if anything is excellent or praiseworthy—think about such things. [9] Whatever you have learned or received or heard from me, or seen in me—put it into practice. And the God of peace will be with you." (Philippians 4:8-9)

Chapter 12

THE POWERS OF HEAVEN AND THE PEOPLES OF THE EARTH

The Bible says: *"All the peoples of the earth are regarded as nothing. He does as he pleases with the powers of heaven and the peoples of the earth. No one can hold back his hand or say to him: "What have you done?"* (Daniel 4:35)

Let us look at the declaration from the passage: *"He does as he pleases with the powers of heaven and the peoples of the earth."*

Powers of heaven: All powers and authority comes from heaven. Jesus Christ said the same thing to Pontius Pilate during His trial. There is absolutely no power anywhere on the earth, except the powers God send to the earth from heaven.

That is why the same Bible passage above says: *"All the peoples of the earth are regarded as nothing."* This statement does not mean that God does not value the people He created and placed on the earth. He values humanity immensely. As a matter of fact, it is how well placed humanity is to God's heart that made God create the earth and everything in it to serve man. The earth was created for man's purpose—man's journey to salvation. Man's journey to salvation goes from Adam to the end of time—

" Come near, you nations, and listen;
  pay attention, you peoples!
Let the earth hear, and all that is in it,
  the world, and all that comes out of it!
2 The LORD is angry with all nations;

his wrath is on all their armies.
He will totally destroy them,
*   he will give them over to slaughter.*
³ Their slain will be thrown out,
*   their dead bodies will stink;*
*   the mountains will be soaked with their blood.*
⁴ All the stars in the sky will be dissolved
*   and the heavens rolled up like a scroll;*
all the starry host will fall
*   like withered leaves from the vine,*
*   like shriveled figs from the fig tree."* (Isaiah 34:1-4)

We are God's special creation, and the whole earth and the universe were created by God to show us what we mean to Him: He made us in His own image and reserved a special place for us in eternity—sons and daughters of the Living God. He filled the whole earth and the universe with mountains of signs for humanity to see His work and be amazed by it. He created man in his own image, giving man His Spirit so that man would be connected to His God forever. But instead, man rebelled against God and cut himself off from God; because he delights in the way of the mind and the pleasures of the world. Unfortunately for man, the spirit must rule the world not the mind. Therefore by choosing his mind over his spirit, man forfeited the authority and power to rule the world Satan who talked him into exercising his mind over his spirit. Adam and Eve realized that their minds were inferior to their spirts; but changing back was not an option. They made the wrong choice and had to live with the consequences of it.

The statement *"All the peoples of the earth are regarded as nothing,"* simply refers to the powerlessness of humanity. Mankind has absolutely no powers. That is why Jesus Christ said to the disciples, *""I am the vine; you are the branches. If you remain in me and I in you, you will bear much fruit; apart from me you can do nothing." (John 15:5)*. And it did not just become that way with Jesus Christ. It was that way with Adam and Eve until they decided to give it all up. It was the way God intended for humanity forever but the mind got in the way.

And this is not an exaggeration. It is the whole truth about man's entire existence. Mankind was created to live in God and operate in God to be able to fulfil its destiny. The scientists that take a swipe at God are simply ignorant. That they are alive and operate at all is God.

There is nothing that God created which He created outside of Himself. The earth and the universe is filled with the Living God. Nothing exists outside of God. That is why the Bible says: *"'For in him we live and move and have our being.' As some of your own poets have said, 'We are his offspring.'"* (*Acts 17:28*)

The problem with the thinkers in our world is that they look at everything they could see and compare God to that. They think of God as though He has any comparison to anything in nature. All of nature can fit into one palm of the Living God. All of the universe could fit into His clenched fist. That is why God said in the Book of Isaiah:

"This is what the LORD says—
the Holy One of Israel, and its Maker:
Concerning things to come,
do you question me about my children,
or give me orders about the work of my hands?
[12] It is I who made the earth
and created mankind on it.
My own hands *stretched out the heavens;*
I marshaled their starry hosts." (Isaiah 45:11-12)

The God of the earth and the universe declared it: **"My own hands** *stretched out the heavens; I marshaled their starry hosts."* (*Isaiah 45:12*) the whole expanse that we call "space" was stretched out in one earth day by the Living God; with his own hands. You Christian teachers and preachers who get cozy with the scientists that are critical of God, what are you thinking.

Anybody that is critical of God would never recognize truth even if it glares into his eyes. Why do you look to him who is rebellious to God for direction? He has gone off in the wrong direction, because he thinks of God as they He is at his level or at the level of anything humanity could conjure up. He can only lead you astray.

The truth that God made you privy to in the Bible is the irrefutable truth. The truth of the Bible is unmatched. You, preacher! Your Bible is a super scientific Book. If you've got a big mind, wrap it around the Bible and watch the truth of God explode before you in scientific terms.

Many people have been misled by the so-called human science. Mankind looks to everything human science currently proposes to validate the gospel! That is a mistake. The gospel validates human science when it is the truth. And the truth in human science confirms the truth of the Bible, because the truth is always the truth.

The gospel has been around way before there was even the slightest hint of science. The world now treats science as the golden rule: if it is not supported by science, then it is not true. You are a preacher, and not an unpaid lab intern! You are a Christian teacher and not an uncelebrated apprentice! The servant of God declares the truth of God boldly and fearlessly because he trusts him the truth.

You get together and modify the truth of the Bible to suit today's political correctness, or worse still, your own purpose! Who are you serving; the living God or the politically correct? The word of God transcends politics and everything else the world's civilization brings into the world.

The Bible currently has a million different translations because everyone believes they got the right angle to the truth. All correct angles to the truth are the same angle. That is the wonderment of God. Christian leaders are letting their minds lead them in spiritual things instead of allowing the Spirit of God to lead them. That is why we all end up with our own disparate flavors of "truth" which is no truth at all.

How can the truth which is one evolve into a million different interpretations? Yet we blame the secular world for thrashing God and religion. Should we not stop and take stock; and see the disservice we the believer have done to the kingdom of God and come back to the truth? The truth is one and indivisible. The truth has only one flavor, and that flavor is Jesus Christ.

The truth of God remains the same: from the time of Adam and Eve to this day. Our God does not change; as such, His truth does not change. He is God and there is no other. And all correct roads to the Living God are the same road or we are all going to different destinations. It is not alright to continue to deepen our differences and hope that somehow we will all converge at the end of the day.

It takes our collective efforts to spread the good news of the gospel around the globe. It will equally take our collective efforts to reunite the church. How can we be sure that we are gathering with Jesus Christ and building with Him when we cannot work with one another?

The Bible says: if we cannot love our brother whom we have seen, how can we love God whom we have not seen? In other words, if we cannot agree and work productively with our brothers whom we interact with regularly in Christian ministries, how can we work with Jesus Christ whose ways we have totally abandoned to follow the dictates of our individual minds? We have all ignored the heart of the teaching of Jesus Christ and are "leaning on our own understanding." That is why everybody's truth differs. Yet we all hope to satisfy God in our ways.

Jesus Christ called on God the Father saying:

""I have revealed you to those whom you gave me out of the world. They were yours; you gave them to me and they have obeyed your word. ⁷ Now they know that everything you have given me comes from you. ⁸ For I gave them the words you gave me and they accepted them. They knew with certainty that I came from you, and they believed that you sent me. ⁹ I pray for them. I am not praying for the world, but for those you have given me, for they are yours. ¹⁰ All I have is yours, and all you have is mine. And glory has come to me through them. ¹¹ I will remain in the world no longer, but they are still in the world, and I am coming to you. Holy Father, protect them by the power of your name, the name you gave me, so that they may be one as we are one. ¹² While I was with them, I protected them and kept them safe by that name you gave me. None has been lost except the one doomed to destruction so that Scripture would be fulfilled.

¹³ "I am coming to you now, but I say these things while I am still in the world, so that they may have the full measure of my joy within them. ¹⁴ I have given them your word and the world has hated them, for they are not of the world any more than I am of the world. ¹⁵ My prayer is not that you take them out of the world but that you protect them from the evil one. ¹⁶ They are not of the world, even as I am not of it. ¹⁷ **Sanctify them by the truth; your word is truth.** *¹⁸ As you sent me into the world, I have sent them into the world. ¹⁹ For them I sanctify myself, that they too may be truly sanctified.*

²⁰ "My prayer is not for them alone. I pray also for those who will believe in me through their message, ²¹ that all of them may be one, Father, just as you are in me and I am in you. May they also be in us so that the world may believe that you have sent me. ²² I have given them the glory that you gave me, that they may be one as we are one— ²³ I in them and you in me— **so that they may be brought to complete unity. Then the world will know that you sent me and have loved them even as you have loved me.**

²⁴ "Father, I want those you have given me to be with me where I am, and to see my glory, the glory you have given me because you loved me before the creation of the world.

²⁵ "Righteous Father, though the world does not know you, I know you, and they know that you have sent me. ²⁶ I have made you known to them, and will continue to make you known in order that the love you have for me may be in them and that I myself may be in them."" *(John 17:6-26)*

Jesus Christ in His supplication to the heavenly Father asked that the Living God made all His disciples to come into complete unity in order to make the world know that God sent Him and loved us as He loved Jesus Christ. Why is this not happening in Christendom today even though Jesus Christ expressly asked the Almighty God for that? We are all unwilling to have God bring us all in complete unity.

Our worldly ambitions have corrupted our minds, making us pursue spiritual things, such as God commanded all His disciples to do, with our minds and not our spirits. We have become our own champions at the expense of the God who gave us His name. We are all serving our own appetites when we should all be serving the Christ who made unbelievable sacrifice at Calvary for all of us.

Branding of the Christian religion has become more important to the Christian leadership than the fellowship of the Holy Spirit of God for the men and women of faith in dire need of the power of God in their lives. The fruits of the Spirit have become rarer than meteorites. We have become more contented with our celebration of life and exclusive membership in the church of God than ensuring that our congregation lives godly lives. The institution that we call the church has become more important that the lives of the suffering men and women that constitute the church.

The command is to come out from them and be separate; and not try and outdo them in their schemes and deceits. We have developed insatiable appetite for the cutting edge technologies, no matter the expense. And our altars have become like concert halls; and our broadcasts, like television productions. We have shaped our outreach to the people to look like Star Track movies. Sound bites and enticing titbits have replaced honest explanation of the message of the cross.

We put more time and expense in the packaging of message than the content and the truthfulness of it. Every preacher these days want to come out on the Television, or at least be part of a Television show, because those who manage to get to the Television have struck it rich. Godliness is not about shows! Godliness is a personal commitment to God and to self. Godliness takes time to perfect.

It is a journey that everyone has to be on. It is a continuing debt that is never fully paid. It is not how much you talk about it, but how often and how readily you apply godliness in every aspect of your life. Godliness has no end, so no matter how much you apply it, you can never exhaust the possibilities to be godly. As you increase in godliness, your grace increases.

So, everybody in the business of spreading the word of God has to become less obsessed with promoting their own brands of the gospel, and readily get with everybody else in the same pursuit; with a single goal of changing souls every single day for the kingdom of God.

How much we tell the world we did is meaningless because none of what we credited to ourselves may mean anything to God. We should focus on the deeds of saving souls; and leave the counting to God who keeps the real counts, and matches it with true rewards. Godliness can never be measured the secular way, or it is not truly godliness.

That we have more poverty in the church today—at the same time we are spending a whole lot more in packaging and transmitting the gospel—is a testament that we are failing and not succeeding in obeying the commands of God.

That the rate of disease and disabilities in the church today is no different from the rate of diseases and disability in the secular world calls to question our loyalty to God and our obedience to His commands. God does not make empty promises. He keeps all His promises.

But these numbers between the church and the secular world may suggest that our God is really not doing the things he promised to do for the church; when in reality, it is the unfaithfulness of the believers that is tarnishing the glory of God and withholding God's grace from the believers. Any religion that fails to lead to godly living is an empty religion and is devoid of the grace of Jesus Christ.

One of the problems diminishing faith in Christendom is that many in the church have subscribed to the human science as much as the secular world—including the parts of science which question the authenticity of the truths God declared in the Bible. All of the claims of Genesis with regard to creation are literal, but many in the church do not believe them because they are convinced that human science got creation right. That shows that many in the church are truly ignorant of the very God they profess:

God created the earth and water before time, and before space and the universe. On Day 1, God dawned light on the earth and time started. Ionization from the light started to produce earthly gases as we know it today, to support life on the earth. On day 2, He stretched out space in readiness for the explosion He employed in creating the sun, the stars, the moon and all the other celestial bodies on Day four.

On Day 3, through unprecedented cataclysmic volcanic eruptions, God made the land came out of the sea, trapping channels of water in the cooling earth as reservoir for rivers and streams. Huge tsunamis clapped over the newly formed land surface, cooling the land and weathering the rocks to create soil for the planting of green vegetation. Hills, mountains and valleys developed as molten land rose unevenly from the bed of the sea, creating ponds and lakes.

Later on Day 3, God planted all the green vegetation over the land He had created. Then on the very next day—Day 4 of creation, the universe was born out of the so-called Big Bang.

I am not using the same terminology as the world scientists for lack of a better term. I am using that term because the explosion that resulted from God's command of Genesis 1:14 rivalled any explosion before that day and ever since. That explosion is what the prophet Isaiah was heralding in the following passage:

"He spreads out the northern skies over empty space;
__he suspends the earth over nothing.
[8] He wraps up the waters in his clouds,
__yet the clouds do not burst under their weight.
[9] He covers the face of the full moon,
__spreading his clouds over it.
[10] He marks out the horizon on the face of the waters
__for a boundary between light and darkness.
[11] **The pillars of the heavens quake,**
__aghast at his rebuke.
[12] By his power he churned up the sea;
by his wisdom he cut Rahab to pieces.
[13] By his breath the skies became fair;
his hand pierced the gliding serpent.
[14] **And these are but the outer fringe of his works."** *(Job 26:7-14)*

And God is promising an implosion of the universe on the Last Day when heavens recede like a scroll:

"I watched as he opened the sixth seal. There was a great earthquake. The sun turned black like sackcloth made of goat hair, the whole moon turned blood red, [13] and the stars in the sky fell to earth, as figs drop from a fig tree when shaken by a strong wind. [14] **The heavens receded like a scroll being rolled up, and every mountain and island was removed from its place.**

[15] Then the kings of the earth, the princes, the generals, the rich, the mighty, and everyone else, both slave and free, hid in caves and among the rocks of the mountains. [16] They called to the mountains and the rocks, "Fall on us and hide us from the face of him who sits on the throne and from the wrath of the Lamb! [17] For the great day of their wrath has come, and who can withstand it?"" (Revelation 6:12-17)

And everything the Bible says about how God created the earth and the universe is all science—super science!

Ifeanyi Chukwujama

Chapter 13

GENERAL WISDOM

Proverbs 22

Thirty Sayings of the Wise

Saying 1

¹⁷ Pay attention and turn your ear to the sayings of the wise;
apply your heart to what I teach,
¹⁸ for it is pleasing when you keep them in your heart
and have all of them ready on your lips.
¹⁹ So that your trust may be in the LORD,
I teach you today, even you.
²⁰ Have I not written thirty sayings for you,
sayings of counsel and knowledge,
²¹ teaching you to be honest and to speak the truth,
so that you bring back truthful reports
to those you serve?

Saying 2

²² Do not exploit the poor because they are poor
and do not crush the needy in court,
²³ for the LORD will take up their case
and will exact life for life.

Saying 3

²⁴ Do not make friends with a hot-tempered person,
do not associate with one easily angered,
²⁵ or you may learn their ways
and get yourself ensnared.

Saying 4

*[26] Do not be one who shakes hands in pledge
    or puts up security for debts;
[27] if you lack the means to pay,
    your very bed will be snatched from under you.*

Saying 5

*[28] Do not move an ancient boundary stone
    set up by your ancestors.*

Saying 6

*[29] Do you see someone skilled in their work?
    They will serve before kings;
    they will not serve before officials of low rank.*

Saying 7

*23 When you sit to dine with a ruler,
    note well what[a] is before you,
[2] and put a knife to your throat
    if you are given to gluttony.
[3] Do not crave his delicacies,
    for that food is deceptive.*

Saying 8

*[4] Do not wear yourself out to get rich;
    do not trust your own cleverness.
[5] Cast but a glance at riches, and they are gone,
    for they will surely sprout wings
    and fly off to the sky like an eagle.*

Saying 9

*[6] Do not eat the food of a begrudging host,
    do not crave his delicacies;
[7] for he is the kind of person*

who is always thinking about the cost.
"Eat and drink," he says to you,
    but his heart is not with you.
[8] *You will vomit up the little you have eaten*
    and will have wasted your compliments.

Saying 10

[9] *Do not speak to fools,*
    for they will scorn your prudent words.

Saying 11

[10] *Do not move an ancient boundary stone*
    or encroach on the fields of the fatherless,
[11] *for their Defender is strong;*
    he will take up their case against you.

Saying 12

[12] *Apply your heart to instruction*
    and your ears to words of knowledge.

Saying 13

[13] *Do not withhold discipline from a child;*
    if you punish them with the rod, they will not die.
[14] *Punish them with the rod*
    and save them from death.

Saying 14

[15] *My son, if your heart is wise,*
    then my heart will be glad indeed;
[16] *my inmost being will rejoice*
    when your lips speak what is right.

Saying 15

[17] *Do not let your heart envy sinners,*
  but always be zealous for the fear of the LORD.
[18] *There is surely a future hope for you,*
  and your hope will not be cut off.

Saying 16

[19] *Listen, my son, and be wise,*
  and set your heart on the right path:
[20] *Do not join those who drink too much wine*
  or gorge themselves on meat,
[21] *for drunkards and gluttons become poor,*
  and drowsiness clothes them in rags.

Saying 17

[22] *Listen to your father, who gave you life,*
  and do not despise your mother when she is old.
[23] *Buy the truth and do not sell it—*
  wisdom, instruction and insight as well.
[24] *The father of a righteous child has great joy;*
  a man who fathers a wise son rejoices in him.
[25] *May your father and mother rejoice;*
  may she who gave you birth be joyful!

Saying 18

[26] *My son, give me your heart*
  and let your eyes delight in my ways,
[27] *for an adulterous woman is a deep pit,*
  and a wayward wife is a narrow well.
[28] *Like a bandit she lies in wait*
  and multiplies the unfaithful among men.

Saying 19

[29] *Who has woe? Who has sorrow?*
  Who has strife? Who has complaints?

Who has needless bruises? Who has bloodshot eyes?
30 Those who linger over wine,
  who go to sample bowls of mixed wine.
31 Do not gaze at wine when it is red,
  when it sparkles in the cup,
  when it goes down smoothly!
32 In the end it bites like a snake
  and poisons like a viper.
33 Your eyes will see strange sights,
  and your mind will imagine confusing things.
34 You will be like one sleeping on the high seas,
  lying on top of the rigging.
35 "They hit me," you will say, "but I'm not hurt!
  They beat me, but I don't feel it!
When will I wake up
  so I can find another drink?"

Saying 20

24 Do not envy the wicked,
  do not desire their company;
2 for their hearts plot violence,
  and their lips talk about making trouble.

Saying 21

3 By wisdom a house is built,
  and through understanding it is established;
4 through knowledge its rooms are filled
  with rare and beautiful treasures.

Saying 22

5 The wise prevail through great power,
  and those who have knowledge muster their strength.
6 Surely you need guidance to wage war,
  and victory is won through many advisers.

Saying 23

[7] *Wisdom is too high for fools;*
in the assembly at the gate they must not open their mouths.

Saying 24

[8] *Whoever plots evil*
will be known as a schemer.
[9] *The schemes of folly are sin,*
and people detest a mocker.

Saying 25

[10] *If you falter in a time of trouble,*
how small is your strength!
[11] *Rescue those being led away to death;*
hold back those staggering toward slaughter.
[12] *If you say, "But we knew nothing about this,"*
does not he who weighs the heart perceive it?
Does not he who guards your life know it?
Will he not repay everyone according to what they have done?

Saying 26

[13] *Eat honey, my son, for it is good;*
honey from the comb is sweet to your taste.
[14] *Know also that wisdom is like honey for you:*
If you find it, there is a future hope for you,
and your hope will not be cut off.

Saying 27

[15] *Do not lurk like a thief near the house of the righteous,*
do not plunder their dwelling place;
[16] *for though the righteous fall seven times, they rise again,*
but the wicked stumble when calamity strikes.

Saying 28

[17] *Do not gloat when your enemy falls;*
  when they stumble, do not let your heart rejoice,
[18] *or the LORD will see and disapprove*
  and turn his wrath away from them.

Saying 29

[19] *Do not fret because of evildoers*
  or be envious of the wicked,
[20] *for the evildoer has no future hope,*
  and the lamp of the wicked will be snuffed out.

Saying 30

[21] *Fear the LORD and the king, my son,*
  and do not join with rebellious officials,
[22] *for those two will send sudden destruction on them,*
  and who knows what calamities they can bring?

Ifeanyi Chukwujama

ABOUT THE AUTHOR

My life is a laboratory. And all human beings are designed as such by the all-knowing God. The only difference among us is that while some willingly become part of life's experiments, some view it from the sidelines.

The best lessons we each learn in life comes to us directly and not through a teacher in an academic setting. We all learn and mature in our experiences by trial and error, just like a scientist in the laboratory. But we are not only the scientist, we are also the test specimen and the laboratory facility & instrumentation—all rolled into one.

And when we are in tune with our spirits, it becomes more verification than 'trial and error' because through our spirits, God feeds us great knowledge about our lives, the things around us and deeper mysteries than we ever thought possible.

Most of my books happened that way. Information came into my mind and takes residence. I soon become aware of it and try to know more about it. As I explore it, it deepens and more is downloaded onto my spirit. And intuitively, I am led to its verification. Once verified, it becomes common knowledge to me.

God has been unbelievably good to me by opening windows to me into great mysterious, such as I have been writing about in my many books. There is hardly a day that I am not writing books. I work on several titles simultaneously, capturing the information while it has not been corrupted either by me or the people I hear on the Television!

Other Titles form this Author:

- What is Love!
- Christ is in Everyone!
- Christianity is not a Religion!
- The Singleness of God!
- Overcoming Your Trials!
- Live the Abundant Life!
- Science, Evolution and God!
- Reflections of Life!
- The Rapture, the Tribulations and the Church!
- The Big Bang: and Jesus Christ birthed the Universe!
- Government is a spirit and the Beast; Science is the False Prophet!
- Revive your marriage instantly be obeying God!
- Scientific Proof that the Earth & Water Existed before Time, Space & the Big Bang!
- Whoever says that sex is good is a liar!

www.ingramcontent.com/pod-product-compliance
Lightning Source LLC
Chambersburg PA
CBHW020910180526
45163CB00007B/2697